World Directory of Pesticide Control Organisations

Third Edition

World Directory of Pesticide Control Organisations

Third Edition

Compiled by

George Ekström
National Chemicals Inspectorate (KemI), Solna, Sweden

THE ROYAL SOCIETY OF CHEMISTRY
Information Services

ISBN 0-85404-437X

A catalogue record for this book is available from the British Library

Published by The Royal Society of Chemistry,
Thomas Graham House, Science Park, Milton Road,
Cambridge CB4 4WF, UK

Typeset by Computape (Pickering) Ltd, Pickering, North Yorkshire
Printed by Hartnolls Ltd, Bodmin, Cornwall, UK

Preface

The first edition of the International Directory of Pesticide Control Organizations was conceived by Professor Bengt von Hofsten and published by the Swedish National Food Administration (Uppsala) in 1979. Between 1983 and 1986 three more editions were published in Sweden. In 1988 the copyright was transferred to the Royal Society of Chemistry, who subsequently published directories in 1989 and 1994, the latter in cooperation with the British Crop Protection Council. This edition, therefore, is the seventh, and the third by the Royal Society of Chemistry.

The purpose of this directory is to encourage international co-operation in the control of pesticides and similar compounds used for protecting food crops in the field and after harvest.

Over the years, pesticides have made a very important contribution towards increasing the quality and quantity of crop production worldwide, and will continue to do so for the foreseeable future. Insects, rodents, nematodes, fungi and bacteria destroy huge quantities of food and feeds, and modern agriculture extensively uses pesticides, fumigants and preservatives to prevent such losses. Agrochemicals are also widely used for weed control and to regulate the growth of food crops.

Despite these obvious advantages in using pesticides, there is concern over the *ever-increasing* use of such chemicals, especially in developing countries. By their very nature, their manufacture, handling and use can involve great personal hazard to those who come into contact with them. Workers involved in the actual application of pesticides are, of course, at particular risk, especially when handling concentrates, but what greatly increases public concern is that the risks from such chemicals extend to the environment generally and to public health by means of residues in agricultural produce and water supplies.

Reasons for the control of pesticides are explained in detail in the following chapters and the role of United Nations agencies and similar organisations in achieving this goal is elucidated.

Note: the compiler of this directory, George Ekström, KemI, P.O. Box 1384, S-171 27 Solna, Sweden, would appreciate any amendments and comments for improvement.

Contents

Intergovernmental Forum on Chemical Safety

Bo Wahlström

THE NATIONAL CHEMICALS INSPECTORATE (KEMI), SOLNA, SWEDEN

Introduction

The Intergovernmental Forum on Chemical Safety is a new mechanism for cooperation among governments for promotion of chemical risk assessment and environmentally sound management of chemicals. The purpose of the Forum is to provide policy guidance with emphasis on regional and subregional co-operation, develop strategies in a coordinated and integrated manner, foster understanding of the issues, and promote the required policy support needed to discharge these functions. The Forum is not a new agency; it is just a mechanism to integrate and consolidate national and international efforts to promote chemical safety.

The concept of the forum was launched in the preparatory process for UNCED, it was recommended by a meeting of government-designated experts held in London in 1991, encouraged by UNCED in Agenda 21, and endorsed by the General Assembly of the United Nations in 1992. The Forum was established by the International Conference on Chemical Safety, held in Stockholm in April 1994 upon the invitation of the Executive Heads of UNEP, ILO and WHO.

The Conference established the Forum and adopted terms of reference for the Forum. It also adopted a Resolution with recommendations on priorities for action. The Conference declared that the Conference should be considered as though it were the first session of the Forum.

Purpose, aims and functions

The Forum is a non-institutional arrangement whereby representatives of governments meet to consider and to provide advice and, where appropriate, make recommendations to governments, international organizations, intergovernmental bodies and nongovernmental organizations involved in chemical aspects of chemical risk assessment and environmentally sound management of chemicals.

The Forum will seek consensus among decision makers representing governments on the development of strategies for the implementation of Chapter 19 of Agenda 21. It shall provide (i) clear and consistent advice for cost-effective, integrated risk assessment and management of chemicals using environmentally sound and generally accepted principles, including advice on alternatives to banned or severely restricted chemicals, and (ii) improved delineation and mutual understanding of roles, initiatives and activities both within and among governments and international organisations having responsibilities for chemical safety.

The functions of the Forum are consultative and advisory, and include a great number of issues such as to identify priorities for cooperative action on chemical safety, recommend concerted international strategies for hazard identification and risk assessment, assist in securing the collaboration of national, regional and international bodies active in the field of chemical safety, advise governments in their work on chemical safety, promote the strengthening of national coordinating mechanisms and of national capabilities and capacities for chemicals management etc.

Participation

There are three groups of participants of the Forum: (i) Government Participants; (ii) Intergovernmental Participants; and (iii) Nongovernmental Participants.

Intergovernmental and Nongovernmental Participants do not have the right to vote.

The representative of each government should be a senior official concerned with chemical safety, including health and environmental issues.

Organization of work

After the inaugural session in Stockholm 1994 two Sessions of the Forum are planned. The next session will take place in February 1997 in Canada, prior to the Special Session of the UN General Assembly in 1997 concerning implementation of Agenda 21. The following is planned for the year 2000, *inter alia* to review progress of Chapter 19. At that time the need for further sessions will be reviewed.

At each session the Forum shall elect from among the representatives of the Government Participants a President, three Vice-Presidents and a Rapporteur. The first Forum elected the following:

President:	Sweden
Vice-Presidents:	China
	Mali
	Mexico
Rapporteur:	Russian Federation

There is also an *Intersessional Group* of the Forum. The Group meets between sessions of the Forum in order to provide advice to the cooperating organizations of the IPCS and to deal with the work of the Forum between sessions. The first meeting was held in Bruges, Belgium, in March 1995 and the second in Canberra, Australia, in March 1996.

The Intersessional Group consists of the officers of the Forum and 26 Government Participants elected by the Forum, taking into account equitable regional and geographical representation. Other participants of the Forum have the right to attend meetings of the Group without the right to vote.

The following countries were elected at the first Forum:

Africa:	*Asia:*
Gabon	China
Kenya	India
Mali	Indonesia
Mauritius	Iran
Nigeria	Japan
Tunisia	Republic of Korea*
	Kuwait

*shall be replaced by Tonga after half of the mandate-time has elapsed

Eastern Europe:	*Latin America:*
Czech Republic	Brazil
Estonia	Colombia
Hungary	Mexico
Russian Federation	Nicaragua
	Trinidad and Tobago

Western European and Other States:
Australia
Belgium
Canada
Germany
Spain
Sweden
Turkey
United Kingdom
United States

The Forum or the Intersessional Group may establish *ad hoc Working Groups* of government and other agreed participants for a set period of time to undertake specific tasks.

A trust fund for the Forum has been established by the Director-General of WHO. Within the framework of IPCS a small Secretariat for the Forum, with an Executive Secretary, has been set up in Geneva.

International Activities to Reduce Releases of Persistent Organic Pollutants (POPs)

Agneta Sundén Byléhn

UNEP – CHEMICALS (IRPTC), GENEVA, SWITZERLAND

Background

Recently a number of international and regional activities have been initiated to reduce and eliminate emissions and releases of persistent organic pollutants (POPs). POPs are chemical substances which are persistent, bioaccumulate and pose a risk of causing adverse effects to human health and the environment. Today, it is widely accepted that the use of such persistent, bioaccumulating and toxic substances cannot be considered a sustainable practice. However, for different social and economic reasons such substances are still in use and/or released to the environment. With the evidence of long-range transport of these substances to regions where they have never been used or produced and the consequent threats they pose to the environment of the whole globe, the international community has now, on several occasions called for urgent international actions to reduce and eliminate releases of these chemicals.

Global Initiatives

At the UNEP governing Council in May 1995 two decisions were taken with regard to global actions on POPs, namely, decision 18/32 which specifically addresses POPs at the global level and decision 18/31 which concerns the protection of the marine environment. A third decision, 18/12, concerning Prior Informed Consent (PIC) and further measures to reduce the risks from hazardous chemicals, also refers to activities to be undertaken under decision 18/32.

An 'International Expert Meeting on Persistent Organic Pollutants: Towards Global Action', jointly organized by Canada and the Philippines in Vancouver in June 1995 brought together experts from academia, industry, international institutions and national authorities from all over the world to discuss the POP-issue on a global level. The meeting provided a series of statements including *inter alia* that domestic regulatory arrangements are not effective in managing the adverse global impacts of POPs, and requested *inter alia* that a suitable international agency (e.g. UNEP) provide definitions, criteria and a comprehensive list of POPs by December 1997.

UNEP Governing Council Decision 18/32 on Persistent Organic Pollutants

Decision 18/32 addressed directly the need for international actions to reduce and eliminate releases and emissions of POPs. It invited the Inter-Organization Programme for the Sound Management of Chemicals (IOMC), together with the International Programme on Chemical Safety (IPCS) and the Intergovernmental Forum on Chemical Safety (IFCS), to initiate an assessment process, beginning with a short list of twelve POPs (aldrin, dieldrin, endrin, chlordane, DDT, heptachlor, hexachlorobenzene, mirex, toxaphene, PCBs, dioxins and furans) already discussed under the LRTAP process of UN/ECE (see below). The decision 18/32 further invited IFCS to develop recommendations and information on international action to be considered at the 1997 sessions of the UNEP Governing Council and the World Health Assembly.

In response to the invitation in decision 18/32 UNEP convened on behalf of IOMC, in October 1995, an *Ad Hoc* Working Group on POPs. The Working Group submitted the workplan to the second meeting of the Inter-Sessional Group (ISG2) of IFCS, held in Canberra, Australia, 5–8 March 1996. ISG2 approved the outline of work and adopted the *Ad Hoc* Working Group.

Within the framework of IOMC, IPCS initiated the assessment work by consolidating existing knowledge on the twelve POPs resulting in an assessment report which was submitted for consideration by ISG2. ISG2 considered that there is a need for some further information for the development of recommendations on international action, and some countries and UNEP offered to do further work to collect and assess such data. ISG2 further agreed that an IFCS-sponsored expert meeting be convened in June 1996 in the Philippines, to assess information on especially existing sources of POPs and the availability of alternative substances and techniques for the substitution and reduction of releases of POPs. This meeting was immediately followed by an open meeting of the *Ad Hoc* Working Group on POPs to develop recommendations and information on international action, including information needed for a possible decision regarding an appropriate international legal mechanism on POPs.

UNEP Governing Council Decision 18/31 on the Protection of the Marine Environment from Land-Based Activities

Decision 18/31 concerns a Global Programme of Action for the Protection of the Marine Environment from Land-Based Activities. The decision specifically addresses POPs in that it calls on States to consider how, within the Global Programme of Action, appropriate attention could be given to support action at national and regional level on POPs. At the Intergovernmental Conference to Adopt the Global Programme of Action, which was held in November 1995, participating representatives from Governments and the European Commission declared their commitment to protect and preserve the marine environment from the impacts of land-based activities, and to do so *inter alia* by acting to develop a global, legally binding instrument for the reduction and/or elimination of emissions, discharges and, where appropriate, the elimination of the manufacture and use of the POPs identified in decision 18/32.

The Global Programme will trigger actions on the national, regional, and international level for the reduction and/or elimination of POPs releases. A strong linkage between the Global Programme of Action for the Protection of Marine Environment form Land-Based Activities and UNEP's initiative on OPs within the IOMC is indispensable.

UNEP Governing Council Decision 18/12 on the Development of an International Legally Binding Instrument for the Application of the Prior Informed Consent Procedure and on Further Measures to Reduce Risks from Hazardous Chemicals.

Decision 18/12 refers first of all to the development of a legally binding instrument for the application of the PIC procedure for certain hazardous chemicals in international trade. In addition, the decision requests the Executive Director of UNEP to convene, in co-operation with FAO, a Government-designated group of experts, to consider what 'further measures' are needed to reduce the risk from a limited number of hazardous chemicals, either within or beyond the scope of Prior Informed Consent, and to report to the Governing Council in 1997. It requests that in considering such further measures work or other relevant fora be taken into account, including IFCS and activities undertaken in accordance with decision 18/32 on POPs. Many of the POP chemicals referred to in Decision 18/32 are also covered by the present voluntary PIC procedure. The work to implement decision 18/12, and especially the development of 'further measures' will hence need to be followed carefully with regard to the development of international actions on POPs.

Regional Initiatives

A number of activities addressing POPs have been initiated at regional level of which some are briefly described below. Among these, one is concerned with long-range transport through air pollution and others are concerned with the protection of oceans and seas. In addition there is one bilateral agreement concerned with the protection of the Great Lakes in North America.

UN Economic Commission for Europe—Convention on Long-Range Transboundary Air Pollution

In 1994, the Executive Body for the Convention on Long-Range Transboundary Air Pollution (LRTAP) under the UN Economic Commission for Europe (UN/ECE) established an Ad Hoc Preparatory Working Group to do preliminary work for a possible Protocol on POPs. The Executive Body agreed in December 1995 that negotiations for a protocol on POPs could be initiated.

Convention for the Protection of the Marine Environment of the North-East Atlantic (OSPAR Convention)

The convention for the Protection of the Marine Environment of the North-East Atlantic was adopted in 1992 by the Ministerial Meeting of the Oslo and Paris Commissions. With this Convention Contracting Parties to the Oslo Convention for the Prevention of Marine Pollution by Dumping from Ships and Aircrafts (1972) and the Paris Convention for the Prevention of Marine Pollution from Land-Based Sources (1974) agreed *inter alia* that discharges and emissions of substances which are toxic, persistent and liable to bioaccumulate, in particular organohalogen substances, and which could reach the marine environment should, regardless of their anthropogenic source, be reduced, by the year 2000, to levels that are not harmful to man or nature with the aim of their elimination. Both the Oslo Convention and the Paris Convention will be superseded by the OSPAR Convention when it enters into force, i.e. after it has been ratified by all Contracting Parties to the Oslo and Paris Conventions.

Barcelona Convention for the Protection of the Mediterranean Sea against Pollution

In June 1995 Contracting Parties to the Barcelona Convention for the Protection of the Mediterranean Sea against Pollution agreed through the Barcelona Resolution to reduce, by the year 2005, discharges and emissions which could reach the marine environment, of substances which are toxic, persistent and liable to bioaccumulate, in particular organohalogens, to levels that are not harmful to man or nature with a view to their gradual elimination. If necessary, reduction measures should be supplemented with programmes aiming at banning the use of such substances. UNEP was entrusted with organizing a consultation meeting with the Contracting Parties, scientific experts, industrialists and NGOs no later than July 1996, in order to facilitate and speed up the definition of methods, programmes and timetables.

The Esbjerg Declaration on the Protection of the North Sea

At the 4th International Conference on the Protection of the North Sea in June 1995 Ministers declared with the Esbjerg Declaration that the objective of ensuring a sustainable, sound and healthy North Sea ecosystem implied the prevention of the pollution of the North Sea by continuously reducing discharges, emissions and losses of hazardous substances thereby moving to the target of their cessation within one generation (25 years). In the context of the Declaration hazardous substances are defined as substances, or groups of substances, that are toxic, persistent and liable to bioaccumulate, whereby toxicity should include chronic effects such as carcinogenicity, mutagenicity and teratogenicity and adverse effects on the function of the endocrine system.

The Great Lakes Agreement

The Great Lakes Water Quality Agreement of 1978 between the United States and Canada, aims at restoring and protecting the Great Lakes. It *inter alia* commits the Governments to virtually eliminate discharges of persistent toxic substances, with 'toxic substance' defined as, 'a substance which can cause death, behavioural abnormalities, cancer, genetic mutations, physiological or reproductive malfunctions or physical deformities in any organism or its offsping, or which can become poisonous after concentration in the food chain or in combination with other substances'.

Note: The Swedish National Chemicals Inspectorate and the Swedish Environmental Protection Agency have published jointly the report 'Alternatives to persistent organic pollutants: the Swedish input to the IFCS Expert Meeting on persistent organic pollutants held in Manila (the Philippines), 17–19 June 1996'. The report can be ordered from KemI, P.O. Box 1384, 17127 Solna, Sweden.

Bilateral Pesticide-related Training Facilities

M. Kern and G. Vaagt

GTZ-PESTICIDE SERVICE PROJECT, POSTFACH 51 80 – D-65726 ESCHBORN, GERMANY

The Food and Agriculture Organisation of the United Nations (FAO) adopted the international Code of Conduct on the Distribution and use of Pesticides in 1985 (amended 1989). This code is the culmination of international efforts to improve user and environmental protection in the use of pesticides.

Together with Chapter 19 of Agenda 21 (UNCED, 1992) both agreements call upon an active role of all involved with pesticides in industrialised and developing countries—governments, industry, traders, users, consumers and non-government organisations—by taking up their responsibilities as defined in the Code of Conduct. The orientation of national technical cooperation towards international agreements serves as an example of a synergistic cooperative effort in this direction between international and bilateral organisations.

As an industrialised nation and major exporter of pesticides, the Federal Republic of Germany has a special responsibility. The German Federal Ministry of Economic Cooperation and Development (BMZ) is therefore committed to support the implementation of both agreements in all its projects. Particular reference is made to the activities of the Pesticide Service Project[1] carried out by the 'Deutsche Gesellschaft für Technische Zusammenarbeit (GTZ) GmbH' (German Agency for Technical Cooperation) on behalf of the BMZ.

The project tasks are directly derived from the Code of Conduct and are directed towards strengthening of government authorities involved in regulation and controlling pesticides. Beside the general information exchange and know-how transfer in pesticide management training is focusing on pesticide residues and formulation control analysis. Since 1994 the Pesticide Service Project is working in this analytical area as WHO-Collaborating Centre for Pesticide Analysis and Training.

The monitoring of the quality and use of pesticides by appropriate laboratories is of major importance in the implementation of pesticide legislation.

The complex field of pesticide analysis place high demands on the

[1] Pesticide Service Project 'Promoting Responsible Conduct', Deutsche Gesellschaft für Technische Zusammenarbeit (GTZ) GmbH, D-65726 Eschborn, Germany, 1993.

laboratory personnel, and on the entire laboratory infrastructure. When laboratories for residue analysis and/or formulation control are being established in developing countires, the Pesticide Service Project provides assistance in the technical planning, in the procurement of the necessary equipment and installations, and in the establishment of analytical procedures. The training capacity is also available to existing laboratories, where technical or analytical problems arise.

Training programmes must be adapted to the specific needs of the laboratory specialisation and situation in the particular country. Courses cover one or more topics as summarised in the table. On-the-job training in other relevant laboratories may also be appropriate.

Table: *Pesticide Analysis Training Courses*

Pesticide residue analysis

- sample processing, clean-up and fundamentals of qualitative and quantitative identification of pesticide residues
- basic course in gas chromatography (GC)
- basic course in high-performance liquid chromatography (HPLC)
- evaluation and interpretation of residue analysis results
- capillary gas chromatography
- troubleshooting in GC/HPLC

Pesticide formulation (quality control) analysis

- sample processing and fundamentals of qualitative and quantitative identification of pesticide active ingredients
- physical and chemical parameters in the quality control of pesticide formulations
- basic course in gas chromatography (GC)
- basic course in high-performance liquid chromatography (HPLC)
- UV/VIS spectrophotometry (IR)
- infrared spectrophotometry (IR)
- troubleshooting in GC/HPLC

If required, other specific topics can also be covered

A major component in all analytical training exercises is the establishment of quality assurance programmes. For this purpose the Pesticide Service Project is conducting every year a ring analysis in residue and formulation analysis.

All quality assurance programmes are embodied in the approach to enhance and ensure the quality of the analytical work of the laboratory and at the same time to give recognition to those laboratories participating in this programme.

Therefore the elements of accreditation or certification schemes, e.g. according to the ISO-guidelines, have become an integral element of all training activities. These quality standards have gained rapidly worldwide attention and recognition and their importance for developing countries is growing constantly.

Further Reading

Promoting Responsible Conduct, GTZ Pesticide Service Project, Eschborn, Germany (1993).

Facts and Figures: A Brief Description of Selected Plant and Post-Harvest Protection Projects, GTZ, Eschborn, Germany (1996).

The GIFAP Safe Use Projects

K. P. Vlahodimus

GIFAP, BRUSSELS, BELGIUM

The need to grow more food safely

With an ever increasing world population, the need to progressively produce more and more food is a stark reality. In 30 years time there will be three mouths to feed for every two today, and it is forecast that more than 95% of those new mouths will be born in developing countries. Countries where food production is frequently constrained due to lack of economic resources which restrict access to modern technology and the more sophisticated and cost-effective farming inputs. Given these restrictions and the need to produce more, it is vital that those farming the land do not, through lack of training, expose themselves to unnecessary risks. This potential danger is recognised in the 'FAO International Code of Conduct on the Distribution and Use of Pesticides'. The Code addresses the need for various groups, including governments, the Crop Protection Industry, trade and international organisations, to work together so that the benefits from the necessary and acceptable use of crop protection products are achieved without adverse effects on people or the environment. The need for improved dissemination of educational materials and information is also recognised.

GIFAP shows the way with Safe Use Projects

The Crop Protection Industry has supported the Code since its introduction in 1985. It has taken proactive steps in developing company stewardship programmes and has increased its commitment to education and training. However, as the FAO acknowledges, the need is for co-operative programmes, involving more than just the Crop Protection Industry. It was with this in mind that GIFAP, the Industry's international trade association, supported by its members, decided to take a positive lead. In 1991, pilot 'Safe Use Projects' were commenced in Guatemala, Kenya and Thailand with the objective of demonstrating that the principles embodied in the FAO Code could be put into practice through a responsible and integrated approach involving Industry, Government, local communities and those working the land. As well as having safety as a key criterion, the projects focused on meeting the increasing demand for food by encouraging the development of sustainable farming systems, including the use of Integrated Pest Management methods, avoiding damage to fragile ecosystems. These projects have demonstrated just what can be achieved and have resulted in the development of numerous training packages and valuable blueprints on how such schemes can be extended into other countries.

Progress

Guatemala

Two thirds of the Guatemalan population (over 9 million, of whom 60% are of Indian origin) live in rural areas. Many children work in the fields and it is estimated that only around 50% of them receive normal schooling. Against such a background, the need for adequate communication and training in the safe use of potentially dangerous products, such as some crop protection products, becomes critical.

Guatemala, in common with the other Safe Use Projects, has achieved some notable successes since 1991. These successes cover the training programmes within the agricultural industry, farmer training, education in schools and medical support services. Interim research in the farming community shows marked improvements in both the knowledge and implementation of crop protection products safety measures both among farmers and farmers' wives who received training in comparison with untrained counterparts.

Key achievements in Guatemala.

- 800 members of the Guatemalan Agricultural Extension Service trained as master trainers in all aspects of integrated pest management, including crop protection products safety.
- Some 475 retailers have been trained and the aim is to have all retailers trained and certificated by 1995 as a prerequisite of continuing to sell crop protection products.
- 226,000 farmers have already received training.
- 2,800 teachers in rural areas have received special training in the safe use of crop protection products and they in turn have trained some 70,000 children. A special educational programme, including audio visual material, based on a scarecrow theme, has been specially developed for schools use.
- Over 2,050 doctors and paramedics have received special training and help is being given in establishing an improved toxicological information service and national data base on poisoning statistics.
- More than 56,000 antidote kits have been distributed.

Kenya

In Kenya, some 85% of the population of 28 million are involved in some way with agriculture, which generates 30% of gross domestic product and accounts for 60% of all exports. However, as only 20% of Kenya's substantial land area is suitable for arable cropping, the first phase of the Safe Use Project was concentrated in the area of intensive agricultural activity south and west of Mount Kenya with future expansion into other key areas, notably in the Western Provinces.

The primary objective of the project was to establish a trainer base and to use the multiplier effect of the extension services to educate 250,000 farmers in the safe handling and use of crop protection products. Prior to this, in mid-1992, independent audit was used to reveal specific shortcomings in handling and use of crop protection products. It also highlighted problems in reading and understanding labels, in communication of precautions by retailers and in the purchase of unauthorised packs. The focus of activity in Kenya was centred on farmer training and an interim audit in 1993 showed encouraging improvements in the attitudes of trained farmers when compared with untrained. More stored crop protection products in locked cabinets; dramatically fewer stored them in the house; pictograms on labels were better understood; used containers were burned or buried; and greater use made of protective clothing.

Achievements in Kenya

- Over 2,000 extension staff were trained as master trainers in crop protection products use and safety by the project team.
- More than 280,000 small scale farmers, primarily in the districts surrounding Mount Kenya, have received training.
- All industry staff involved have been trained in safety and received a certificate of competence—all future staff will have to be trained before beginning their job. Approximately 2,800 retailers have received training and the plan is to have 2,000 accredited by the end of 1995.
- A specific radio programme sponsored by the Safe Use Project has turned into the Kenya Broadcasting Corporation's most successful programme in its 40 year history.
- Following consultation with Project personnel, the Kenyan Government is amending its Pest Control Products Act to improve registration procedures in compliance with the FAO Code and has recruited more inspectors to enforce legislation.

Thailand

Agriculture which involves some 75% of the population, forms an important part of the Thai economy. Here again, partnership with Government and an independent audit, that was used to identify shortcomings, proved to be key factors in establishing the basis of a successful project. Industry developed a checklist of minimum standards relating to plant location, storage, packaging and labelling, employee safety, transport and disposal. Government on its part, through the Department of Agricultural Extension, made a commitment to training for all levels of extension workers, retailers and farmers based on the GIFAP training manuals. They also made available a supporting budget. The high level of co-operation made a very broadly based project, which extended beyond safe use into improved industry standards, possible.

A further development in Thailand is the establishment of a model farm programme supported by the Thai Ministry of Agriculture, the GTZ (German Aid Organisation) and the Safe Use Project. Initial results directed at improving spraying techniques showed a significant reduction in the volume sprayed per hectare, an increase in worker's productivity and a considerable reduction in risk exposure.

Interim audits conducted by the Government's Office of Agricultural Economics, showed significant improvements in the use and availability of protective clothing and in handling and storage practices. However, they also highlighted more needed to be done in the area of rinsing and disposing of empty containers and that problems of repackaging still exist with retailers.

Achievements in Thailand

- Some 2,200 master trainers have now been trained.
- Over 600 retailers have been trained and in future all retailers will have to undergo training and become accredited before being permitted to sell crop protection products.
- 450,000 farmers have been given practical training on a large range of topics.
- 253 schools and 65,000 students have received training on the importance of safety.
- Some 600 hospitals and 1,300 doctors and paramedics have received training on the diagnosis and treatment of crop protection products poisoning.
- Economical sets ($4) of protective clothing have been made widely available.

Future needs and perspectives

The pilot projects established in Guatemala, Kenya and Thailand have each in their own way demonstrated just what can be achieved through an integrated effort to raise standards of safety in all aspects of crop protection products' use and application. GIFAP has, on behalf of the Industry, taken a positive lead. It has shown that it is possible for Government, Industry and others to work together with local communities to improve training, education, communication and the levels of expertise and help available from a range of sources, including agricultural extension services, crop protection products retailers and all levels of medical staff.

A valuable resource has been established both in terms of training programmes and the know-how and expertise that has been acquired on how to set up and introduce these programmes into countries located in major areas of need around the world. Similar needs exist in many other countries and as the knowledge of what has been achieved in the pilot countries is disseminated, there is no shortage of demand. There is, however, one major problem—the problem of resources.

Resources to meet the demand

The task is clearly one that is too large for the Crop Protection Industry alone to meet. The unfortunate reality is that in those countries where the need is greatest, the opportunities for the providers of modern technology and inputs are limited, but the need to increase food production is high. However, there is a pressing need for all concerned with the well being of rural economies, and ensuring they have an adequate, safe and affordable supply of food, to join together in extending the implementation of Safe Use Projects. Integration has been shown to be the key to success, but Governments and international agencies have vital roles to play in ensuring that the resources are made available to meet the need. By working with the Crop Protection Industry and utilising the programmes it has developed, much can be done to ensure modern technology can be used in a safe and controlled way so that the benefits can be realised without the risks. Only by working together in a common cause, can standards be raised for the benefit of rural communities throughout the developing world.

Integration leads to success

As the projects developed in Guatemala, Kenya and Thailand, chosen as representatives of developing countries of Latin America, Africa and Asia, the value of integrated involvement as the vital ingredient of success, became progressively more apparent. Although the Crop Protection Industry led the way, it needed the full co-operation of government, national and local, in its role as controller of health, education and agriculture staff and services; of international agencies; the agricultural distribution chain; and last but not least, local communities and farm owners and workers.

The task that faced the three experienced agronomists that were appointed in 1991 to head up the then embryonic projects, was enormous. Their subsequent achievements highlighted the key elements of success.

Key elements of a successful project

- Careful auditing and assessment of all aspects of each market to enable realistic objectives and targets to be set.
- Winning the commitment and support of government and the local agricultural industry.
- Formulating detailed plans.
- Preparing comprehensive training packages—once the necessary level of support had been obtained.

- Identify and train the trainers who would be drawn from different backgrounds depending on the groups they were to train.

If the projects were to succeed, they had to be comprehensive and include all the links in the various chains of involvement.

Who has been involved ?

- Crop Protection Industry
  - ⇒ Factory workers
  - ⇒ Distribution staff
  - ⇒ Salesmen and technical advisors
  - ⇒ Retailers
- Farming
  - ⇒ Farmers
  - ⇒ Farm workers
- Medical
  - ⇒ Hospital staff
  - ⇒ Doctors
  - ⇒ Paramedics
- Education
  - ⇒ Teachers
  - ⇒ Children
  - ⇒ Groups of wives of farmers and farm workers
  - ⇒ The media

The Future

Industry has now accumulated and organised in a modular form, the knowledge gained from the three pilot Safe Use Projects. This knowledge is gradually incorporated in the portfolio of services provided by the National Crop Protection Associations which are taking initiatives to promote the safe use of crop protection products. This service will be available to support relevant projects undertaken by national authorities, regional services of international organisations and NGOs.

In parallel, Industry's project is extended to priority countries where it is felt that the prevailing risks are particularly high or the local conditions are convenient for its smooth introduction. Thus, during 1996 a number of countries in the developing world will be running their own safe use project. For details on these developments and the safe use project in general, queries should be addressed to the following addresses:

Dr. J.L. Aston
GIFAP Safe Use
P.O. Box 30321
Nairobi
Kenya
Tel. +254 2 86 06 67
Fax +254 2 80 31 72

Mr. W.W. Ellis
GIFAP Thailand
Room 1405, Floor 14,
Rasa Tower Building
555 Pahonyothin Road,
Chatuchak
Bangkok 10900
Thailand
Tel. +662 937 0487 90
Fax +662 937 04 91

Ir. A. Garcia
AGREQUIMA-GIFAP
6A Avenida y 12 Calle,
Zona 10
Edificio 'Punto 10'
Tercer Nivel, Oficina 3D
Guatemala City
Guatemala
Tel. +502 2 32 92 86
Fax +502 2 32 92 88

Appropriate Technology for Pesticide Analysis in Developing Countries

Malin Åkerblom

SWEDISH UNIVERSITY OF AGRICULTURAL SCIENCES, DEPARTMENT OF ENVIRONMENTAL ASSESSMENT, UPPSALA, SWEDEN

John R. Cox

NATURAL RESOURCES INSTITUTE, UNIVERSITY OF GREENWICH, CHATHAM MARITIME, CHATHAM, KENT, UK

Introduction

Modern pesticides in today's agriculture play an important role in the struggle to increase food production, protect harvested food, and to control vectors of disease. A thorough knowledge of the properties of the hundreds of different pesticides, and of their behaviour in the environment in which they will be applied, is the basis for appropriate choice and handling of these substances. There is a continuous need for reliable data concerning residues in food and water, in working areas and in the environment, in order to set priorities for actions to remedy and prevent pollution and to determine the effectiveness of the control actions. Additionally, there is the need for quality control of formulated products, particularly from local formulators and after prolonged storage, and the identification of pesticides in poorly labelled containers or in obsolete stocks.

Analytical laboratories with the ability or capacity to perform pesticide analyses are scarce in many developing countries, and totally lacking in some. This is partly due to the fact that today's published analytical methods increasingly rely on complicated and expensive equipment, such as gas chromatography (GC), high-performance liquid chromatography (HPLC), and mass spectrometry (MS). These techniques make big demands on laboratory facilities and require easily accessible instrumental service facilities, easy access to spare parts and a stable and continuous power supply, demands which are difficult to meet in countries with a fragile infrastructure.

There are, however, alternatives to these analytical methods, which make it possible to perform pesticide analyses in places lacking the basic requirements necessary for sophisticated instrumentation. In this chapter, the principles of such methods are discussed. These methods, although sometimes less sensitive, will still be appropriate to tackle the most pronounced problems frequently met in developing countries. Such problems include excessive residues in food e.g. food sold in the local market too soon after treatment with pesticide; water contaminated due to cleaning of equipment or from spillage and drainage of concentrate and spray solutions; fish kills; leakage to the environment from pesticide stores and obsolete stocks; operator exposure; badly formulated or stored products or products in unlabelled containers. The analytical methods presented here can also be used in field trials concerning dissipation rates of pesticides from crops and the environment, and in occupational health studies. Experienced people can use the methods for checking food items for compliance with maximum residue limits (MRLs).

Overview of pesticide analysis

Pesticide analysis consists of a chain of steps commencing with the initial sampling, followed by sample handling and storage, sample preparation, extraction, clean-up of extract, final determination and reporting. The individual steps are shortly presented here and are more thoroughly discussed later in this chapter.

The very first step is to clearly define the purpose of the study, because it has a bearing on each of the other steps—what to sample, what analytical sensitivity is required, how should the data be reported, etc.

The quality of the data produced throughout the study is extremely important since they will form the basis for decisions and actions. Basic requirements must be fulfilled to avoid both false positive and false negative results, including the calibration of laboratory equipment and instruments, effective maintenance of pesticide reference standards, and the full and regular validation of the analytical method(s) regarding interferences, recovery of pesticides added, lower detection and quantification limits, linearity. All stages of the analysis from sampling through to the final reporting of results must be properly documented.

It is crucial to the analysis that the sample is properly selected, transported and preserved in order to obtain a relevant, representative sample, worthy of analysis. Too often the analysis is wasted because insufficient attention has been paid to these criteria and the validity of the sample is open to question.

Pesticides are extracted from the sample by use of the appropriate, carefully chosen solvent(s). Liquid samples, e.g. water, can also be extracted by percolating through an appropriate sorbent which retains the pesticides for subsequent elution.

During the extraction process, many natural compounds will be extracted along with the pesticides. Clean-up is used to separate the pesticides from these co-extractives. Clean-up procedures vary, each employing different chemical and physical properties to selectively remove the wanted or unwanted compounds e.g. differences in solubility are used in liquid-liquid partitioning, and different affinities for sorbents such as silica and alumina are used in adsorption column or thin layer chromatographic clean-up. Pesticides can also be separated from compounds with higher molecular weight, such as fats and pigments, by size exclusion/gel permeation chromatography (SEC/GPC).

The final determinative step usually involves a form of chromatography and includes both a separation of substances and their selective detection and quantitative measurement, allowing several pesticides to be determined simultaneously. Systems for analytical chromatography include thin-layer chromatography (TLC), gas chromatography (GC), and high-performance liquid chromatography (HPLC). With each of these techniques different separation and detection systems are used for confirmation of results.

Spectrophotometry can be used for many pesticides, e.g. paraquat, benomyl, and dithiocarbamates, and colorimetric kits are available for cholinesterase-inhibiting insecticides and some fungicides.

Immunoassay, which is based on antibody-antigen reaction, can be used for the study of single pesticides or a group of closely related pesticides. The technique is selective and sufficiently sensitive to allow water samples to be analysed without extraction, clean-up and further separation. With certain immunoassay kits some substances can also be determined in crude extracts of other matrices. However, at present its specificity precludes its use in multi-residue analysis, i.e. where several pesticides are determined simultaneously.

The raw data must be evaluated with great care and sound judgement—there are many opportunities to misinterpret data and although experience helps to minimise such errors, experienced analysts are also not immune. Finally, the data should be reported in a manner which is both easy to understand and which cannot be misinterpreted. This is of particular importance where the customer is a non-chemist.

Establishment and Maintenance of a Pesticide Laboratory: Some Basic Considerations

The reasons for establishing a pesticide laboratory vary—e.g. to support research, to monitor residues in food or the environment, or the quality of formulated products. Technologies and procedures used by the different laboratories depend essentially on the function of the laboratories, and these can be poles apart in terms of the complexity and breadth of the techniques used and the equipment required.

A larger, central laboratory often stands a better chance of success than a number of smaller laboratories, and allows for the development of a core of expertise. Such a laboratory is often placed within a Ministry of Agriculture, which is responsible for the safe use of pesticides, needs formulation analysis, knows what and how pesticides are—or should be—used, performs residue dissipation trials etc. In reality what matters is not to which ministry the laboratory belongs, but that its mandate is clear—who it will serve, what the priorities are, and so on. A central laboratory also requires good scientific back-up perhaps through association with a university, but continuity may be better safeguarded if it is not directly controlled by the university. Conversely, small laboratories conducting research on pesticides may be more appropriate to a university.

Pesticide residue analysis is one of the most demanding forms of chemical analysis. It is rare, however, for people not involved in the analyses to understand the efforts needed to produce correct results, and the need for skill, time, and effective troubleshooting etc. Also, without the continuing political desire (and funding) to maintain a functioning pesticide laboratory, a laboratory, if ever established, will soon deteriorate.

The networking of laboratories, within a country or region, would provide valuable support. Analytical results often have to be verified by other techniques, which might be available at some laboratories but not at others. Also, intercalibration, where several laboratories analyse the same sample, is a vital part of the quality assurance concept.

Staff. A pesticide laboratory needs skilled chemists. Even in a small laboratory it is desirable to have at least two chemists, to assist each other, discuss problems and methodology, and for safety reasons. The quality of the analysts involved in pesticide analysis is of vital importance. Analysts must be graduates with a thorough understanding of analytical chemistry, and have an introductory training of a minimum of two to six months, depending on future tasks, in a competent laboratory. Even then, learning an analytical procedure in a manner akin to teaching a parrot how to talk, is of little value. The analyst must get a true understanding of what role each component of a procedure plays, to be able to adapt methods to address problems that arise, and to appropriately validate that revised method. Unfortunately, too few employers allow the time for the analyst to practice this thinking or to properly research the techniques. Furthermore, technicians must be carefully selected to meet the demands, in order not to do more harm than good in the laboratory. They must get proper training, and be given a thorough understanding of why they are performing the different tasks. Finally, management has to be aware that well-trained staff are extremely valuable, and that lack of reward and motivation may lead to their loss to other employment.

Laboratory. There are some basic requirements for the laboratory in terms of accommodation and services. Activities must be well-separated to minimise the risk of contamination. For example there must be the facilities to handle reference standards and samples with high concentrations of pesticides in a room separate from the main laboratory. The analytical equipment must be kept in a clean, non-dusty area, away from sample preparation and, preferably, in a separate room. Good washing-up facilities, separate for residue or concentrate work, are of primary importance. Stable power supply is required, especially for equipment like GC or HPLC, as well as high pressure water supplies. Well-functioning fume cupboards are essential in sample and standard preparation areas. Where the building does not permit fume pipes, a fume cupboard with exchangeable active carbon filter may be used.

Equipment. Much of the basic equipment is common to most laboratories—ovens; balances; water purification system; flask shaker etc. Facilities for extraction and clean-up of extracts include a homogenizer (which could be of household type); centrifuge; rotary or Kuderna-Danish evaporator. A soxhlet extraction system and an ultrasonic bath are optional. Where water pressures are low, a water circulation/cooling facility will be required. Refrigerators and freezers are necessary for the storage of residue samples and extracts. A separate refrigerator must be used for the storage of reference substances.

The type of equipment that is needed for the final determination must be thoroughly considered. In this chapter low-cost, low-tech methods like TLC are emphasized, which however need skilled chemists. Modern analytical procedures published in manuals and scientific journals are increasingly using more sophisticated GC and HPLC systems for high sensitivity/high resolution. The problems faced in developing countries in adopting these techniques are immense and are often not considered at all, particularly since the procedures are generally written by analysts in developed countries, who have not been exposed to the need to undertake analysis in circumstances less than ideal and are often not aware of the problems. GCs and HPLCs are available with greatly varying degree of sophistication. In countries with limited budgets and where there is restricted, or delayed, access to service facilities, the following key questions should be raised before considering purchasing a GC or HPLC: What analyses do we want to perform? What are the options? Are local or regional service facilities available for the make of equipment? At what cost? What are the consumable costs likely to be? For GC, is gas available of the right purity? Are there means to obtain reliable, continuous power supply? What alternatives (other techniques) are there? Do we have somebody who could devote much time to make the instrument work? The list is almost endless, but too often it is overlooked. Cost-conscious analysts need to address these issues carefully, particularly so for those in developing countries with a host of practical, let alone financial, constraints. The less extra features there are in the equipment, the easier it is to maintain. Particular attention must be given to alternatives techniques, at least as a back-up facility.

There is the need for a rigid, continuous programme of laboratory maintenance which does not only cover the repair of obviously broken items, but the checking, calibration and maintenance of all laboratory equipment (including volumetric glassware). The Network of Users of Scientific Equipment in Southern Africa (NUSESA, see references) is an example of networking on techniques, equipment and spare parts. They also compile lists of equipment and people with skills to operate and maintain equipment.

Economics. The costs of establishing the laboratory includes costs for basic facilities, equipment, glassware and chemicals. Assuming a basic laboratory is available, still a fume cupboard ($ 5000), separate fridges and freezers, and good washing facilities may have to be purchased. For a small laboratory

$ 10 000 may be sufficient for basic equipment for extraction and clean-up, such as water purification system, evaporator, centrifuge, ultrasonic bath or shaking machine. Equipment for the final determination varies from $ 3000 for simple TLC system, through $ 30 000–60 000 for each GC or HPLC, to $ 80 000–150 000 for an MS-detector. Basic glassware for a small laboratory would cost $ 7000–10 000 if starting from scratch, with the cost of chemicals being about the same.

Running costs have to be taken into account in the long term planning. These include not only staff salaries and chemicals (including solvents and gases), but also glassware—new items or replacements for what has been broken, instrument consumables, spare parts and service—costs which are substantial for complicated equipment and perhaps 15–25% of the initial price, literature, training courses and conferences, etc.

Sampling and Sample Preparation

Non-chemists are most often involved in the sampling, and it is crucial that they are informed beforehand on how to sample. This section may serve as a guide.

Sampling can in most cases be performed without special equipment, but this does not mean that the step is any the less important; the importance of the sampling step cannot be over-emphasised and well-planned and well-performed sampling is the basis for a reliable analysis. The sample must be taken according to the purpose of the study, e.g. whether the study is for general screening, to support a particular field trial, or aims at finding the worst-case situation such as the highest concentration or the most polluted/contaminated sample. It must be taken such that it represents as far as possible the whole of the unit to be studied. The degree of replication of sampling also needs to be determined at the outset.

A representative sample is obtained by drawing several samples of equal size from the unit to be studied and then thoroughly mixing them. From this a subsample is drawn which is further chopped or ground and mixed in the laboratory before the analytical portion is taken (Kratochvil & Peak, 1989). Glass bottles with glass stoppers or teflon or aluminium lined lids are generally used as sample containers. Sometimes the samples may be wrapped in alumina foil or kept in clean paper bags. Polyethylene bags may be used as an outer envelope but should not be in contact with the sample.

Sample cleanliness is vital. Samples may be easily contaminated in the field if the tools, sample containers and protective equipment used for sampling are not well cleaned, or where the sample is taken in close proximity to a freshly treated area. Pesticides may be lost by degradation, adsorption and, sometimes, by volatilisation from the sample. As a general rule, samples should be kept in dark, air-tight containers and placed, as soon as possible, into cold storage, using ice or freezing blocks, or a transportable (camping) refrigerator or freezer. In the laboratory samples should be stored frozen if not extracted directly upon receipt. With certain types of sample, however, such as root crops, and some fruits and vegetables, freezing destroys the cell structure and subsequent sub-sampling becomes more difficult. In such cases, samples should be mixed and sub-sampled before freezing.

Sampling in markets or in stores is relatively straightforward. With other types of sampling there are potential hazards which must be considered. These include noxious samples or sampling areas, dangerous animals and vectors of diseases, slippery stones or deep holes in a river/lake, etc. Boots, long-sleeved gloves, and other protective garments should be used, as appropriate.

Sampling of different types of media

Food and crop. Food may be bought in the shop or market just as people do for household use. There are international guidelines concerning sampling for pesticide residue in produce in international trade, covering the size and number of items to take, which part of the food items to analyse, etc. (Codex, see references). It is advisable to follow these guidelines if the objective of the study permits. Samples from field trials are taken from several replicate plots, using a carefully defined, statistically valid, sampling plan. Field sampling guidelines are available in the literature or from international bodies and national registration authorities. In the laboratory, clean, sharp knifes and a chopping-board are required for cutting, mixing and subsampling. Household blendors are useful for homogenising fruit, vegetables, meat, and fish, cereals, nuts and pulses, before extraction.

Water is most often collected as a 'grab' sample, in which a single sample is taken at a specific site. It may also be taken as a composite sample of several grab samples, in space or time, which are combined before extraction. Where possible, the sample should be taken directly in the sample bottle to avoid the risk of contamination. If the water is filtered prior to extraction, it must be taken into account that hydrophobic pesticides are adsorbed on suspended solids and will stay on the filter, which may also need to be analysed. Transportation of aqueous samples is a problem and solid-phase extraction systems, in which the pesticides are absorbed from the water onto a sorbent (and which allows sample extraction in the field, see Extraction section) are becoming particularly popular.

Sediment. In many cases it is possible to sample directly into a wide mouth sample bottle with the help of a spoon, closing the bottle before it is taken up from the bottom. Bottom sediment samples are generally collected from the top 2–5 cm. Mechanical samplers are also available for different types of bottom sediment, conditions etc. Before analysis, any 'free' water should be carefully removed without losing any of the fines. Stones and debris, which would not be able to pass a 2 mm sieve, are removed. The results will be given on a dry weight basis, and therefore the moisture content of the sample should be determined.

Soil. Samples are best collected with a soil core sampler or a spade.

Sampling should follow a prepared schedule depending upon the objectives of the sampling and with the desired degree of replication. One scheme is to take samples along an imaginary line diagonal to furrows, or along imaginary S-shaped or zig-zag lines. When using a spade, a hole should be made, a little deeper than the desired depth. A slice about 5 cm thick is taken along the vertical wall of the hole, and from this the desired portion is removed. The sample can also be taken with a spoon directly from the vertical wall of the hole at the desired depth. If a tube sampler (corer) is used, great care must be taken not to contaminate lower levels of the soil profile. Samples should be well mixed in a clean bucket before subsampling. Before analysis stones and debris, which would not be able to pass a 2 mm sieve, are removed. As with sediments, the results are generally given on a dry weight basis, and the moisture content of each sample should be determined.

Fish and other biota may be caught by any method—except poisoning—at the chosen monitoring sites. For fish, it is preferable to sample the same species (both carnivores and herbivores) and sex, and roughly the same size, throughout a monitoring programme. When analysing pesticides that accumulate in the lipid (fat) tissue, it is usual to express the results on a lipid basis. Therefore the lipid content of the sample should be determined.

Exposure studies. A feasible way to estimate occupational exposure to pesticides is by assessment of skin exposure, which is of particular importance since most uptake of pesticides, by field workers, is generally via the skin. Clean cotton pads are attached to different sites of the body, over and under clothing (WHO, 1982); solvent-cleaned cotton gloves may be used for the hands. These sample materials are subsequently extracted with solvent for chemical analysis. Visual assessment, by the incorporation of a fluorescent or coloured dye to the pesticide solution before spraying can be used as an alternative. Air sampling is normally accommodated by sucking air through a filter or a sorbent, by means of a portable pump.

Uptake of cholinesterase inhibitors, i.e. organophosphorus and carbamate insecticides, may be followed by determination of the cholinesterase level in finger tip blood. In the field special kits may be used (WHO, 1984; Vandekar, 1980; McConnell *et al.*, 1992), or the blood can be sampled on a filter paper, which is sent to the laboratory for spectrophotometric analysis (Kolmodin-Hedman & Eriksson, 1987; Augustinsson & Holmstedt, 1965). Urine is useful for measurement of uptake of paraquat (Åkerblom, 1990).

Extraction

Pesticide extraction systems generally revolve around soaking, shaking, blending, ultrasonication or soxhleting with an organic solvent or solvent mixture and are well documented in the scientific literature. The choice of extracting medium is crucial; it must be able to penetrate the matrix, to compete with the analyte for the sorptive sites of the matrix, and to dissolve the pesticide at relevant concentrations. A polar organic solvent, possibly in combination with a nonpolar organic solvent, most often fulfils this requirement. A prior rehydration of matrices with low moisture content to open up cell structures is generally considered essential. For comparison of extraction procedures, samples with field incurred residues must be used.

With an appropriate extracting medium, the choice of the extraction method generally depends upon the speed required for the analysis and the equipment available. Mere soaking of a well homogenised sample, overnight, after initial shaking or ultrasonication is sufficient in most cases. Blending with solvent is generally the quickest procedure (2–6 minutes) but requires an explosion-proof blendor; by comparison soxhlet extraction, which also requires special glassware, can take 4–8 hours, For solid samples, matrix solid-phase dispersion (MSPD) is a technique gaining renewed interest. In this, the sample is ground with sodium sulfate for dehydration and a solid material such as sand or a sorbent that disrupts tissue structure and disperses the sample as a thin film on the material, making it easily accessible to the extraction solvent. The mixture is most conveniently poured into a column and eluted with solvent.

One problem is that these methods require a fairly large amount of solvent. Analytical samples are generally in the range of 25–50 g and the amount of solvent 2–5 times more. However, miniaturisation as described under the section on clean-up is not really an option—to preserve the integrity of the analytical sample it must be representative of the original field sample and little compromise can be made, unless the samples are extremely well-mixed and homogenized, or they are liquid. Depending on the demands of the final determination, solvents of fairly low purity may be purchased and used after redistillation. However, each redistilled batch has to be tested before use for interference etc. This procedure can result in some savings on consumable costs, but should not compromise the quality of the analysis. The recent techniques of supercritical fluid extraction (SFE) and accelerated solvent extraction (ASE) use little solvent but they rely on expensive and demanding equipment.

Solid phase extraction is a genuine solvent-saving technique, often used for the extraction of water samples instead of liquid-liquid extraction with organic solvents. The sample is drawn through a small column of 1–2 g of sorbent, usually modified silicas with chemically bonded C18- and C8-carbon chains (RP-C18 or RP-C8), or a polyvinylstyrene matrix. The pesticides are then eluted from the column with 2–10 ml of solvent. Some systems rely on electrical or water suction pumps (positive or negative pressure) to facilitate the movement of water through the column. However, if the sorbent material has a low level of small particle size material, of 'fines', hand pressurising a reservoir of the sample (with a rubber bulb) will allow an adequate flow. Ready-made commercial cartridges are normally used, but the use of bulk material in laboratory packed column will help reduce the costs. These systems allow for field collection and extraction of water samples with minor equipment requirements.

Clean-up

The need for clean-up of sample extracts is dependent on many factors such as type of sample, concentration levels targeted, and the degree of selectivity of the extraction procedure and the final determinative step. A wide range of procedures has been developed to separate pesticides from co-extractives, and many apply to a separate group of pesticides. Basically they reflect three main techniques: liquid-liquid partitioning, column chromatography and chemical treatment. These may be used alone, or in combination for very impure extracts. Although some of the techniques normally use special equipment, this can be circumvented, and solvent can be saved by scaling down traditional methods.

The differing solubilities of pesticides and co-extractives in two immiscible solvents is used in liquid-liquid partioning, most often performed in separatory funnels; pH-changes are used to alter the partition characteristics of ionizable compounds. Often traditional partionings can be scaled down to test-tube separations to save solvent, although at this level increased care must be taken to minimise the risk of error.

The column chromatographic techniques most frequently used are adsorption and gel-permeation chromatography. Gel-permeation chromatography (GPC, more correctly called size-exclusion chromatography, SEC) is used for removal of lipids, pigments, humic acids and other large molecules from sample extracts, whereas adsorption columns separate compounds mainly according to polarity.

Typical materials for column adsorption chromatography are alumina, florisil, silica gel and charcoal/celite/magnesium oxide mixtures, eluted by gravity in measured fractions with solvents/solvent mixtures of increasing polarity. Relevant procedures are well-documented in the literature. In the past these methods have, however, tended to use relatively large quantities of adsorbent (10–40 g) with correspondingly large volumes of eluting solvents. The advent of solid phase extraction systems (SPE)—commercially available cartridges packed with between 0.2 and 1 g of adsorbent—allowed for separations to be made with greatly reduced solvent volumes. Elution needs vacuum or pressure, most simply provided by a glass syringe. Although the cartridges are relatively expensive ($ 3–5 each), this is partly outweighed by the reduced costs on solvents. The packing materials are separately available, and with experience they can be used for laboratory-packed columns, which again reduces the overall cost.

The philosophy of scaling down the clean-up has been extended by many laboratories to traditional column packings, although there is relatively little published (Specht and Tillkes, 1985). These miniaturised procedures tend to use columns of between 0.5 and 1.0 cm i.d. with between 0.2 and 1.0 g of packing. Typical elution volumes range from zero (the extract is poured through with the eluate being directly collected and analysed) to about 10 ml. Useful packings include 40–75 5µ silica gel or RP-C18 column bulk material, dry-packed in a pasteur pipette to about 4 cm. Pressure from a dropper bulb lightly held to the 'column' will speed the flow. With these miniaturised techniques, there can be greater scope for error or between-run variation and care must be taken. Miniaturisation should not be introduced without the appropriate validation.

Thin-layer chromatography (TLC) is also well suited to adsorption chromatographic clean-up, with an identified portion of the eluted plate being removed and extracted for subsequent analysis.

Hydrophobic GPC/SEC provides a, virtually, universal clean-up for pesticide analysis. It is applicable to hundreds of pesticides and a wide range of matrices, although it is particularly useful for the clean-up of extracts of high fat content. A benefit of the system is that all contaminants are discharged from the column enabling it to be continuously re-used until performance deteriorates. Generally several hundred clean-ups can be made with a single column before it needs replacement. This enables the unit cost of the analysis to be dramatically decreased. GPC systems are commercially marketed as pumped, automatic systems. It can be equally well applied using a semi-automated system, if an HPLC pump and injector is available, or as a fully manual system using gravity feed or gentle pressure applied by an air stream (could be provided by an aquarium pump). Early systems used about 30 g of packing, (usually a Bio-Beads SX-3 packing, 200–400 mesh) with elution volumes of up to 250 ml. However, as with other column chromatographic systems, these can also be scaled down. Systems using between 2.0 and 8.0 g of Bio Beads SX-3 (removes 0.2–0.4 g of fat), and eluting with 10–50 ml 1:1 ethyl acetate/cyclohexane are used by the authors (Cox, 1995; Åkerblom, 1995). These systems can be operated in either pumped or gravity fed mode and produce significant savings in consumable materials; with gravity fed systems, the analysis time can be slightly extended although the efficiency of the separation and the sample recovery are unaffected.

Chemical clean-up treatments are generally specific and not used in multi-residue analysis. However, they are useful in certain circumstances and particularly so for chlorinated pesticides. For example, when a hexane extract is gently mixed with concentrated sulfuric acid for half a minute, most of the co-extractives are destroyed whereas most organochlorine pesticides are unaffected. Dieldrin and endrin are exceptions, but for them warm ethanolic potassium hydroxide can be used in a similar way. (Jensen *et al.*, 1983; Åkerblom, 1995).

Sweep co-distillation, in which pesticides are separated from, particularly, fatty extracts by volatilisation and cold-trapping is another technique that has been used by many workers and which, under carefully defined conditions, can produce some excellent results. Again, however, it is not a multi-residue procedure but of value where the target pesticides are known and have been shown to be amenable to this form of separation. Although equipment and gas is needed, the dependence on solvents and reagents is reduced.

Thin-Layer Chromatography (TLC)

Thin layer chromatography can be used for the final determinative step. It is relatively low-cost technique appropriate for many routine screening operations as well as for confirmation of the identity of suspected pesticides. It has its limitations with regard to powers of separation and sensitivity and these vary between individual pesticides. Still, TLC offers the opportunity to undertake analysis where other, more sophisticated, facilities are not available, or for those compounds that are not easily analysed with other techniques. With an experienced operator, certain pesticides can give responses almost comparable in sensitivity, precision and accuracy to GC or HPLC methods.

Thin-layer chromatography uses the principles of liquid chromatography (cf. Clean-up section). The adsorbent is coated on a uniform flat surface, such as on one side of a plate of glass. A row of spots of sample extracts and reference standard solutions is applied on the layer some distance from one edge of the plate. This edge is dipped into the mobile phase, generally an organic solvent mixture, and the solvent moves up the plate, by capillary action, and across the spots; this is termed plate development. Components with different adsorption characteristics move along the plate at different rates, effecting the separation. Once the plate has been developed for a suitable period of time, the presence of any pesticides is highlighted using an appropriate visualisation system.

Silica and alumina, with a particle size of up to 50 μm, are most commonly used for pesticides. The layers for analytical work are usually 0.1–0.25 mm thick, but may be up to 2 mm if used for preparative work such as clean-up. There is a wide range of ready-made plates commercially available, with the sorbents fixed to glass or alumina sheets. The layers can also be prepared in the laboratory by spreading a slurry of adsorbent, in water, across glass plates and allowing it to dry. Ready-made plates are advantageous as they are easier to handle, are more uniform, and have both smaller particles and a narrower particle size distribution than laboratory-made plates, thus giving better separations and quantifications. They can be cut into smaller plates to reduce costs when e.g. testing an eluent. With ready-made plates reproducibility is increased although some may have binders which may interfere with the detection procedure chosen, which has to be tested, and which is avoided with laboratory-made plates. Laboratory-made ones are considerably cheaper, if the spreading equipment is available and once the skill to make them is developed.

Sample extracts and standard solutions, generally in the volume range 5-550 μl, are applied to the sorbent layer by capillary tube or hypodermic syringe. A typical sample load is an amount of cleaned-up extract corresponding to 0.2 to 1 g of a vegetable sample. A stream of cold air from a hair dryer or an aquarium pump evaporates the solvent and speeds up the procedure.

After the plate has been developed with the selected solvent or solvent mixture, it is sprayed with, or dipped into, a visualisation reagent, unless the reagent has already been incorporated into the layer. Spraying devices may be driven by compressed air (a rubber bulb or an aquarium pump can be used) or by a propellent gas.

There are numerous selective visualisation systems in the literature (e.g. Sherma, 1973; Ambrus, 1981; Cox, 1995). Some examples are presented below:

Halogenated compounds, such as the organochlorines and halogenated pyrethroids, phenoxy acids and triazines, may be detected in quantities of 5–50 nanograms with silver nitrate either incorporated in the layer or sprayed on the plate. After development the plate is exposed to intense UV light from a lamp source in a shielded box. Halogenated compounds react with the silver nitrate to form greyish spots of colloidal silver against the white or faintly brown background.

Organophosphorus and carbamate insecticides are detected by enzymatic reaction, some in nanogram quantities or less. The plate is sprayed with a buffered solution containing a cholinesterase source, such as serum from horse or pig, cow, pig or chicken liver homogenate, extract of heads of bees or flies, or commercial cholinesterase. The enzyme is inhibited where these pesticides are present. The plate is then sprayed with a substrate such as acetylthiocholine, naphthylacetate or indoxylacetate. Intact enzyme hydrolyses the substrate, giving acetic acid. A pH indicator is sprayed (not necessary with indoxylacetate) to indicate where no acid has been produced, which is where the cholinesterase was inhibited by the pesticide.

Triazine and urea herbicides are visible down to nanogram quantities after spraying with a mixture of a redox indicator and a preparation of chloroplasts, easily obtained by grinding green leaves of e.g. wheat or beans in a mortar.

Fungicides are detected in amounts of 20 to 100 ng after spraying with a suspension of spores of *Aspergillus niger* (grown on agar in petri dishes). After one or two days at 37 °C the spores will have grown on the plate except where fungicide is present.

For the quality control of formulated products commercial plates with a UV fluorescent indicator incorporated are well suited. Non-fluorescent compounds appear as dark spots when viewed under UV light. This visualisation system is more general than the others, which allows screening for many different compounds including transformation and degradation products of parent pesticide, and which is acceptable when there are few interfering co-extractives. It is less sensitive than the above detection systems, but with formulation analysis there is little restriction on quantities of sample applied to the plate.

To identify the pesticides, the spots of the samples are compared with those of the standards, run on the same plate. A result obtained in one chromatographic system is only an indication of identity, and the sample has to be run in another system to improve certainty of identity. There are tables in the literature indicating the Rf-values, i.e. the migration distance of the compound vs that of the solvent front, for different systems (e.g. Sherma, 1973; Ambrus, 1981; Åkerblom, 1995). A complex mixture is best run in two-dimensional systems, where the sample, spotted in one corner of the plate, is first run in one solvent or solvent mixture and, after turning the plate 90°, in another. Standards are spotted in the neighbouring corners, each eluted in only one solvent system, for comparison. The colour of the spot can also be used for identification purposes, as well as a different spray reagent, and for instance spectra or immunoassays of spots extracted from the plate.

Quantification is by comparing the size and intensity of the spot of the sample extract with those of standard solutions. The best results are obtained with densitometers, but there are means to get good enough results without this equipment. Visual estimation can achieve accuracy and reproducibility in the 10–30% range, and this degree of accuracy is adequate in most situations in residue analysis. For more accurate results, sample and standard spots are applied alternating in such a manner that the concentration of sample increases from one side of the origin line to the other while the amount of standard increases in the opposite direction. After detection, two neighbouring standard spots bracketing a sample spot are used for estimation.

The quantification achieved in this way is suitable for residue analysis where sampling often reflects a compromise between feasibility and statistical requirements. This means that results obtained are seldom valid for more than one significant figure, nor does the biological significance warrant more figures. Therefore the one-figure data obtained by TLC are not inferior to those obtained by GC or HPLC. In fact, many draw wrong conclusions from the second (or multiple) figure(s) where given.

In recent years, interest in TLC has increased in industrialised countries in parallel with the development of different instruments to improve its capabilities. Laboratories can, however, start with simple manual methods and gradually upgrade to more sophisticated ones using high-quality commercial plates, an automatic sample spotting device to give better uniformity of spots, and a UV/VIS scanner for improved quantification and lowered limit of detection. Ultimately, the laboratory can invest in an apparatus for high-performance thin-layer chromatography (HPTLC) or for automatic multiple development (AMD; for manual application see Lodi *et al.*, 1991) for excellent separation of spots. All this can also be combined with selective colour reactions for different groups of pesticides. Although these more refined techniques are vulnerable to some of the problems of GC and HPLC, when problems develop, the analyst can revert to more manual methods of operation and informative results can still be obtained.

Gas and High Performance Liquid Chromatography

The emphasis in this chapter has been laid on thin layer chromatography as an important analytical tool for analysts in developing countries. By contrast, GC and HPLC will only be briefly reviewed here.

These instrumental techniques probably account for well over 90% of all pesticide residue analyses, with Gas-Liquid Chromatography (GLC or GC) generally being the preferred option. For laboratories with large scale multi-residue programmes analysing hundreds of pesticides, GC and HPLC will often be used in combination to accommodate the required broad range of pesticides. This equipment is expensive and, surprisingly often considering the cost, the wrong selection is made by the analyst at the time of purchase. Before purchasing, the key questions given earlier (section on Establishment of a laboratory) must be asked.

Gas-Liquid Chromatography

For GC the pesticides must be volatile at elevated temperatures. Very polar or ionized substances must first be converted to volatile derivatives. In principle, a gas chromatograph comprises an injector, a column and a detector, all kept at elevated temperatures, e.g. for pesticide analysis with an injector at between 185 and 250 °C, a column at between 90 and 280 °C and with a detector at between 250 and 350 °C. A few microlitres of the sample solution are injected through a membrane into the injector, the substances vaporise and are carried through the column by a carrier gas. The column can be either a glass tube with an internal diameter of 2–4 mm and 1.5–2 m long and filled with packing material, or a capillary tube of fused silica, 15–60 m long with an internal diameter of 0.2–0.5 mm. A non-volatile organic liquid, the stationary phase, is attached to the particles of packed columns or to the walls of capillary columns. The passage time through the column is dependent both on the volatility of the compound and its affinity for the stationary phase. The time spent in the column (retention time, RT) is characteristic of each pesticide under a specific set of operating conditions. When the substances emerge from the column, they are detected in the detector, which sends signals, proportional to the amount of substance, to a recorder, and each chemical is registered as a peak on the chromatogram, i.e. the printed record of the chromatographic analysis.

The choice of column system is dependent on the analytical requirements, experience of the staff and available funding. Capillary columns are expensive ($ 400–800) but give very good separations if handled correctly. Among the capillaries the megabore columns (0.53 mm i.d.) are more robust towards impure extracts and overload, and need less sophisticated injection systems. Packed columns are an attractive and cheaper alternative and there is much published retention time data to support their use (e.g. Zweig vol VI, 1972; Thier & Zeumer, 1987; Åkerblom, 1995). Packed columns are easy to handle and can be repacked in the laboratory with many different stationary phases. Since many compounds (including co-extractives) may have the same or similar retention times, the identity of a compound observed must always be confirmed by running the sample on at least one more column, with a different stationary phase.

It is strongly recommended, both for safety reasons, security of supply and for convenience, to purchase gas generators instead of using gas cylinders.

The generators give a steady supply of pure gases and although initially rather expensive to buy (\$ 7000–12 000), they may be more economical in the long run.

GC can employ a wide range of possible detection systems: flame ionisation (FID, generally for the analysis of formulated products—few residue applications); flame photometric (FPD, for sulfur and phosphorus containing compounds); thermionic or nitrogen-phosphorus (TID or NPD, for nitrogen and phosphorus containing compounds); electron capture (ECD, generally for compounds with high electron affinities e.g. halogenated compounds); Hall's conductivity detector (for halogenated compounds); mass selective detector (ITD or MS).

Laboratories select their GC requirements depending upon the likely balance of pesticides to be analysed, but typically the laboratory may need a minimum of two or three detector types. Increasingly laboratories involved in routine produce analysis are using GC-MS as a one-stage analysis, with its ability of identifying compounds on the basis of molecular ions and unique fragmentation patterns. However, the MS is expensive and requires very skilled personnel and excellent basic facilities.

For chromatogram generation, a chart recorder is quite sufficient. Integrators can save time on peak area measurement and in the generation of retention time data although systems tend to be increasingly more computerised, with the accompanying risk of problems, particularly where remote from service facilities.

High-performance liquid chromatography, HPLC

The setup of HPLC is similar to that of GC, but the chromatography is carried out at ambient or slightly elevated temperature with solvent as the carrier (mobile phase), and UV absorption is most often used for detection.

HPLC is of limited value in routine multi-residue analysis because of the relatively poor response of the UV detector to many pesticides and its lack of selectivity. Poor response can, in part, be compensated for by the higher sample loadings that can be achieved, and HPLC is used routinely for the residue analysis of some pesticides and, particularly, some herbicides, fungicides and other non-volatile or thermally labile compounds. Paraquat, diquat and benomyl are often analysed by HPLC, but they can equally well be determined by spectrophotometry (see section on Colorimetry). Rodenticides is another area of application, substituting TLC.

HPLC is the best choice for the quality control of formulated products since it gives better quantitative results than GC and is suitable for non-volatile compounds as well.

The columns are generally of stainless steel, 15–30 cm long and with an i.d. of 3–5 mm and are delivered ready-packed, mostly with silica or modified silicas of 4–10 μm particle size. Narrow bore cartridge columns are also finding increasing favour amongst analysts as these are often more convenient to install, cheaper to purchase and tend to use smaller volumes of mobile phase, thereby reducing laboratory consumable costs. The small particle size of the column packings together with a very narrow particle size distribution, is the basis for the good separation achieved. These fine particles, however, create a high back pressure and the mobile phase must be driven by a pump or a system of pumps, able to give a constant flow at high pressure (100–300 atm).

Fixed wavelength UV detectors, or diode array detectors which are able to provide full UV spectra to give selectivity, are generally used. Fluorescence can be used for certain compounds, notably for insecticidal carbamates in combination with a post-column derivatising unit. Mass selective detectors are also available but are still too demanding for most laboratories.

Immunoassay

Immunoassays are analytical techniques based on the highly selective binding of an analyte to an antibody. They are rapidly gaining in popularity for environmental analysis because of their simplicity, speed and sensitivity. They can readily be adapted to on-site testing methods, and do not need expensive and complicated equipment. The analyses are performed with test kits which contain all that is needed for a semi-quantitative measurement, including detailed instructions for their use.

The assay is performed either on plates containing a set of pre-treated microtitre wells, or in pre-treated test tubes. Measured quantities of sample and reagent are added to the plate or tube and, after a suitable reaction time, a new set of reagents is added, and a colour developed. Concentrations are estimated by visually comparing sample and standard with regard to the colour developed in the test. A special photometer is available for more precise quantification.

For obvious reasons the tests are not suited for multi-residue studies, but are advantageous when screening for single substances or sometimes groups of substances. The immunoassay can also be combined with other techniques, e.g. to verify a TLC or HPLC indication of a substance by eluting the TLC spot or sampling the HPLC fraction, and testing it with the assay.

Most kits for pesticides are developed for the analysis of water, and are then sensitive down to 0.02–2 μg/l, depending on substance. In many cases the kits can be used for other matrices after extraction, some clean-up, and transfer to a polar solvent diluted with water. Applications include ground and surface water monitoring; rapid field or screening tests; worker re-entry and spray drift monitoring; enforcement and inspection of foodstuffs.

The kits are quite expensive to buy ($ 300–1000 for 96 tests). However, the cost per sample is rather low (even allowing for samples to be run in triplicate, the usual practice) compared with the cost of chemical residue analysis, all costs included, when analysing single compounds. The kits should be stored in the refrigerator and have a limited shelf life.

As with all analytical methods, there is the risk for false positive and, especially for coloured samples, false negative results. Thus, as always, pesticide-free samples and samples with known quantities of pesticides added should be analysed in parallel; significant positive findings should be verified with other techniques.

Colorimetric Analysis

Colorimetry is well known as an analytical technique with a pattern of use across most analytical disciplines. A spectrophotometer is therefore often already available for use, either in one's own laboratory or in another laboratory in the neighbourhood. Several methods were developed for the analysis of single pesticides or groups of pesticides before chromatographic systems became more widely used (Zweig & Sherma, see references). Many procedures for e.g. carbamates (Appaiah *et al.*, 1983), benomyl/carbendazim (Steurbaut *et al.*, 1973), thiabendazole (Wood, 1976) and diquat, paraquat (Kirsten, 1966, Åkerblom, 1990) are still in use in parallel with GC or HPLC techniques, or for confirmative purposes. UV spectra for identification and quantification can also be obtained after a TLC clean-up.

Modern spectrophotometers allow increased sensitivity of response for some of the older methods, increasing their value to the analyst. Also, when scanning a spectrum in the second derivative mode, selectivity is enhanced and sensitivity may increase 5–50 times. Sensitivity can be further increased by using longer path-length sample cuvettes.

Spectrophotometric procedures for the quality control of formulated products are available (CIPAC; Zweig & Sherma, see references).

Test kits for colorimetric analysis, based on a discrete chemical reaction, are usually intended for use in the field and do not require a pesticide laboratory. They are few in number and most have been developed to support specific studies without a more general use in mind. An example is a field test kit, based on a simple dye reaction, for the determination of the concentration of the fungicides imazalil and thiabendazole in banana-treatment solutions and as a qualitative test to determine whether or not the fruit has been treated (Cox, 1995). Likewise, several commercial colorimetric kits for the enzymatic determination of cholinesterase inhibiting pesticides are available (Cos, 1993). The use of kits was also described by Soong and Lo (1989) and Blewett and Krieger (1990). A review on cholinesterase determination as a diagnostic tool for detecting human exposure to pesticides has been conducted by Vandekar (1980), and a colorimetric kit for use in the field was tested by McConnell *et al.* (1992).

The kits are usually semiquantitative, and can be less sensitive than other analytical techniques. They are generally intended for single pesticides or groups of pesticides, and are best suited for situations where the identity of the pesticide is known. The use of the kits must be carefully considered and designed, and the results obtained must be critically reviewed. However, under well-defined circumstances such kits offer a relatively cheap means of analysis.

Conclusions

Although pesticide analysis, and especially the analysis of residues, is demanding, it can with care and within defined limits be performed in remote places with a low level of infrastructure, lack of reliable electricity, and low budgets.

The main requisite is not sophisticated analytical instrumentation but the indigenous and dedicated analyst, with a thorough understanding of the philosophy of pesticide analysis. Most types of analyses can then be performed without access to GC or HPLC, equipment which is used in most pesticide laboratories and which need stable power supply and easy access to the appropriate service facilities and spare parts.

Many studies on pesticide residues can be performed with thin layer chromatography (TLC), immunoassays, or colorimetric methods. These techniques are especially suited for studies on a single or a few pesticides, such as field dissipation or exposure studies. TLC is also suited for the identification of pesticides from obsolete stocks or those found to be leakaging from formulation plants or old stores. TLC can still be used for multi-residue screening of pesticides in samples with unknown history, e.g. food in the market, especially in the local market, since in a given area rarely more than a dozen different pesticides may be used by smallholder farmers. For import and export control, when looking for a much wider range of pesticides, then GC is better suited.

Although the sensitivities obtained with these simpler methods in many cases match those of GC and HPLC, this is not generally the case. However, the levels reached by more sophisticated equipment are rarely biologically relevant and are not necessary for the primary demands in a developing country.

For a laboratory purchasing an instrument like GC or HPLC there are a host of key questions that must be posed and answered before any decision is made, e.g. on the type of analysis to be performed, availability of service facilities, running costs and so on. In most cases, the less sophisticated, the more likely it is that the instrument will continue working for several years.

However, no matter which technology is used, the basic requirements for the correct sampling, the validation of methods, effective record keeping, laboratory cleanliness and other good laboratory practices remain the same and can never be corner-cut.

In conclusion, a well designed and performed study can be conducted without access to sophisticated equipment. The most important element is for a well-trained and dedicated analyst, with strong and continual managerial support.

General references

Åkerblom, M. (1995) Environmental monitoring of pesticide residues. Guidelines for the SADC region. Monitoring techniques series Vol 5. Southern Africa Development Community, Environment and Land Management Sector, P.O. Box 24, Maseru 100, Leshoto.

Analytical Methods for Pesticide Residues in Foods. Health and Welfare, Canada. Canadian Government Publishing Centre, Supply and Services, Ottawa, Canada, K1A 0S9. ISBN 0–660–12213–8.

Analytical Methods for Residues of Pesticides in Foodstuffs. Government Publishing Office, Cristoffel Plaintijnstraat 2, Postbus 20014, Gravenhage, Netherlands. ISBN 90–12–04672–6.

AOAC, Association of Oficial Analytical Chemists. Official Methods of Analysis; new edition every 5 years. Available from AOAC, 2200 Wilson Blvd., Suite 400–FF, Arlington, VA 22201–3301 USA. (Validated analytical methods for the determination of constituents and contaminants, including pesticies)

Bauer K.; Gros L.; Sauer W. (1991) Thin layer chromatography—An introduction. Hüthig Buch Verlag GmbH. Heidelberg. Available from Merck, Germany

CIPAC Handbook, Volumes 1, 1A, 1B, 1C, 1D and addenda as published. Collaborative International Pesticides Analytical Council Limited. Heffer and Sons, Cambridge, UK. (Validated analytical methods for formulated products.)

Cox J. R. (1995) Pesticide residue analysis. Training manual. Natural Resources Institute, Central Avenue, Chatham Maritime, Chatham, Kent ME4 4TB, UK.

Cox, J. R.; Halliday, D.; Kilminster, K. (1990) A guide to the establishment and maintenance of pesticide laboratories in developing countries. Natural Resources Institute, Bulletin no. 28, 1990. Available from Publications and Publicity Section, Natural Resources Institute, Central Avenue, Chatham Maritime, Chatham, Kent ME4 4TB, UK.

Deutsche Forschungsgemeinschaft/The International Association of Forensic Toxicologists (1992) TLC Rf values of toxicologically relevant substances on standardized systems. VCH Verlagsgesellschaft, Weinheim. ISBN 3–527–27397–2.

Food and Drug Administration Pesticide Analytical Manual. US Department of Health and Human Services, Public Health Service, Food and Drug Administration, Washington, DC, USA.

Miller, G. (1992). Pesticide residue analysis in the food control laboratory. Manuals on food quality control, 13. FAO Food and Nutrition Paper 14/13. ISBN 92–5–103269–6. Available from Publications Division, FAO, Viale delle Terme di Caracalla, 1–00100, Rome, Italy.

Sherma, J.; Fried, B. eds. (1991) Handbook of thin-layer chromatography, Marcel Dekker, Inc. New York & Basel, ISBN 0–8247–8335–2.

Stahl, E. (1969) Thin-Layer Chromatography. Springer Verlag, Berlin, Germany.

Thier, H.-P.; Zeumer, H. eds. (1987) Manual of pesticide residue analysis. Deutsche Forschungsgemeinschaft. Vol 1. ISBN 3–527–27010–8 VCH Verlagsgesellschaft, Weinheim.

Thier, H.-P.; Kirchhoff, J., eds. (1992) Manual of pesticide residue analysis. Deutsche Forschungsgemeinschaft. Vol. 2. ISBN 3–527–27017–5. VCH Verlagsgesellschaft, Weinheim. Available from VCH Verlagsgesellschaft, P.O. Box 101161, D–6940 Weinheim, Germany.

Tomlin, C. ed. (1995) The Pesticide Manual. 10th ed. ISBN 0–948404–79–5. Available from British Crop Protection Council Publications Sales, Bear Farm, Binfield, Bracknell, Berkshire RG12 5QE, UK. (A handbook on chemical, physical, toxicologic and environmental properties of each of over seven hundred pesticides.)

Zweig, G. and/or Sherma, J., eds. Analytical Methods for Pesticides and Plant Growth Regulators. Volume I: Principles, Methods and General Applications, 1963; Volume II: Insecticides, 1964; Volume III: Fungicides, Nematicides and Soil Fumigants, Rodenticides, and Food and Feed Additives, 1964; Volume IV: Herbicides, 1964; Volume V: Additional Principles and Methods of Analysis, 1967: Volume VI: Gas Chromatographic Analysis, 1972; Volume VII: Thin-Layer and Liquid Chromatography and Analyses of Pesticides of International Importance, 1973; Volume VIII: Government Regulations, Pheromone Analysis, Additional Pesticides, 1976; Volume IX: Spectroscopic Methods of Analysis, 1977; Volume X: New and Updated Methods, 1978, ISBN 0–12–784311–6; Volume XII: High Performance Liquid Chromatography of Pesticides, 1982, ISBN 0–12–784312–4; Volume XIII: Synthetic Pyrethroids and Other Pesticides, 1984, ISBN 0–12–784313–2; Volume XIV: Modern Analytical Techniques, 1986, ISBN 0–12–784314–0; Volume XV: Principles, Statistics and Applications, 1986, ISBN 0–12–784315–9; Volume XVI: Specific Applications, 1988, ISBN 0–12–784316–7; Volume XVII: Advanced Analytical Techniques, 1989, ISBN 0–12–784317–5. Academic Press Ltd., U.K.

CD-ROM and on-line data retrieval systems are available from, e.g., The Royal Society of Chemistry, Thomas Graham House, Science Park, Milton Road, Cambridge, CB4 4WF, UK; Commonwealth Agricultural Bureaux International, Wallingford, Oxfordshire, OX10 8DE, UK.

Specific references

Åkerblom, M. (1990) Rapid determination of paraquat in urine with ion-pair extraction and spectrophotometry. Bull. Environ. Contam. Toxicol. 45, 157–164.

Anbrus, A.; Hargitai, É.; Károly, G.; Fülöp, A.; Lantos, J. (1981) General method for determination of pesticide residues in samples of plant origin, soil and water. II. Thin layer chromatographic determination. *J. Assoc. Off. Anal. Chem.* **64**, 743–48.

Appaiah, K.M.; Kapur, O.; Nagaraja, K.V. (1983) Colorimetric determination of propoxur and its residues in vegetables. J. Assoc. Off. Anal. Chem. 66, 105–107.

Augustinsson, K.-B.; Holmstedt, B. (1965) Determination of cholinesterase in blood samples dried on filter-paper and its practical application. *Scand. J. Clin. Lab. Invest.* **17**, 573–583.

Blewett, T.C., Krieger, R.I. (1990) Field leaf test kit for rapid determination of dislodgeable foliar residues of organophosphate and n-methyl carbamate insecticides. Bull. Environ. Contam. Toxicol., 45, 120–124.

Codex: Guide to Codex Recommendations Concerning Pesticide Residues: Part 5—Recommended method of sampling for the determination of pesticide residues (Ref.: CAC/PR 5); Part 6—Portion of commodities to which Codex Maximum Residue Limits apply and which is analysed (Ref.: CAC/PR 6). Part 7—Codex Guidelines on good practice in pesticide residue analysis. Available from The Codex Alimentarius Commission, Food and Agriculture Organisation of the United Nations, Via delle Terme di Caracalla, 00100 Rome, Italy.

Cox, J.R. (1993) The role of test kits in pesticide residue analysis, Conference proceedings, Postharvest News and Information 4, 23–24n.

Jensen, S.; Reutergårdh, L.; Jansson, B. (1983) Manual of Methods in Aquatic Environment Research. Part 9. Analyses of metals and organochlorines in fish. FAO Fisheries Technical Paper No. 212 FIRI/T212, p. 21–33.

Kirsten, W.J. (1966) The determination of diquat residues in potato tubers. *Analyst* **91**, 732–738.

Kolmodin-Hedman, B.; Eriksson, K. (1987) Biological monitoring of exposure to organophosphates. A field study using cholinesterase estimation of whole blood dried on filter paper. In: Occupational and chemical hazards. Cellular and biochemical indicies for monitoring toxicity. Foà, ed. Horwood Publisher, 529–533.

Kratochvil B., Peak J. (1989) Sampling techniques for pesticide analysis. In: Sherma J. (ed.) Pesticides and Plant Growth Regulators, Academic Press, N.Y., Vol. 17, 1–33.

Lodi, G.: Betti, A.; Kahie, Y.D.; Mahamed, A.M. (1991) Multiple-development high-performance thin-layer chromatography of organochlorine pesticides. *J. Chromatog*. **545**, 214–218.

McConnell, R.; Cedillo, L.; Keifer, M.; Palomo, M.R. (1992) Monitoring organophosphate insecticide-exposed workers for cholinesterase depression. New technology for office or field use. *J. Occup. Med.* **34**, 34–37.

Network of Users of Scientific Equipment in Southern Africa (NUSESA). Information on NUSESA is obtainable from International Foundation for Science, Grevturegatan 19, S–11438 Stockholm, Sweden.

Sherma, J. (1973) Thin-layer chromatography: Recent advances. In: Zweig (ed.) Analytical Methods for Pesticides and Plant Growth Regulators, Academic Press, N.Y., Vol. 7, 3–87.

Specht, W., Tillkes M. (1985) Gas-chromatographic determination of pesticide residues after cleanup by gel permeation and min-silica column chromatography. Part 5. Cleanup of foods and feeds of vegetable and animal origin for multi-residue analysis of fat-soluble and water-soluble pesticides. Fresenius Z. Anal. Chem. 322, 443–455 (in German: English version in DGF, Vol. 1, 383–400; cited in Åkerblom, 1995).

Steurbaut, W., Dejonckheere, W., Kips, R.H. (1973) Fate of benomyl on early autumn and spring lettuce. Meded. Fac. Landbouwwetensch., Rijksuniv. Gent. 38, 917–928.

Thier, H.-P., Zeumer, H. eds. (1987) Manual of pesticide residue analysis. Deutsche Forschungsgemeinschaft. Vol. 1. VCH Verlagsgesellschaft, Weinheim, 383–400.

Vandekar, M. (1980) Minimizing occupational exposure to pesticides: Cholinesterase determination and organophosphorus poisoning. Residue Reviews 75, 67–80.

WHO (1982) Field surveys of exposure to pesticides. Standard Protocol, WHO/VBC/82.1, WHO, Geneva.

Who (1984) Spectrophotometric kit for measuring cholinesterase activity. WHO/VBC/84.888, WHO, Geneva.

Wood, J.S. (1976). Mertect®. In: Zweig and Sherma (eds.). Analytical Methods for Pesticides and Plant Growth Regulators, Academic Press, N.Y., Vol. 8, 299–308.

International Organisations and Programmes

Hamish Kidd
ROYAL SOCIETY OF CHEMISTRY, CAMBRIDGE, UK

Achim Halpaap
UNITED NATIONS INSTITUTE FOR TRAINING AND RESEARCH, GENEVA, SWITZERLAND

UNITED NATIONS AGENCIES

Food and Agriculture Organisation of The United Nations (FAO)

Via delle Terme di Caracalla
I-00100 ROME
Italy
Tel: (6) 57971
Telex: 610181 FAO

Liaison Offices:

UN Economic Commission for Africa
Joint ECA/FAO Agriculture Division (JAFR)
Africa Hall
P.O. Box 3001
ADDIS ABABA
Ethiopia
Tel: 156834

United Nations Liaison, Geneva (LGEN)
Office of the FAO Representative to the UN Organisations in Geneva
Palais des Nations
CH-12211 GENEVA 10
Switzerland
Tel: (22) 346011, 310211
Telex: 289375

The compilation of information for this chapter has benefitted, inter alia, from the *Background Paper on International Pesticide Risk Reduction Activities* prepared by Kennan Garvey, US EPA, for the OECD/FAO Pesticide Risk Reduction Workshop, Uppsala, Sweden, October 1995, and the *UNITAR Resource Guide on Training and Technical Assistance Activities of International Organizations in the Area of Chemical Management and Safety*, April 1994.

UN Economic and Social Commission for Western Asia
Joint ESCWA/FAO Agriculture Division (JNEA)
P.O. Box 27
BAGHDAD
Iraq
Tel: (1) 5569400
Telex: 2133013 UNECWA

FAO Liaison Office with the United Nations (LUNO)
United Nations Headquarters
1 United Nations Plaza
Suite DC1–1125
NEW YORK, NY 10017
USA
Tel: (Centrex) (212) 754–6036, 754–6039, 754–6041/2/3
Telex: 236350

Aims/Activities

FAO is actively involved in helping countries establish and strengthen their plant protection capabilities, advising on organisation and management, providing operational guidelines and training scientific and technical personnel. This also includes strengthening food control through laboratory assessment, training in laboratory management and residue analysis, assistance in Good Laboratory Practice (GLP) and Analytical Quality Assurance (AQA) activities and publication of manuals for pesticide analysis and quality assurance in food control chemical laboratories.

The risks associated with pesticide use in developing countries are multifaceted and there is no single effective approach to their resolution. In order to be effective decisions must be taken at the national level taking into account the prevailing local economic, environmental and administrative conditions. There are a number of programmes and initiatives in place and under development aimed at helping countries to develop the capability to manage pesticides effectively, which in turn results in a reduction in the associated risks. These activities include training in the safe, efficient and environmentally sound management of pesticides, establishing laboratory infrastructures for quality control of pesticides and residues, as well as the disposal of obsolete and unwanted pesticides.

International Code of Conduct on the Distribution and Use of Pesticides. The International Code of Conduct was developed to address a number of difficulties associated with the use of pesticides in developing countries where adequate regulatory infrastructures are frequently lacking. The Code was adopted by FAO member countries at the 23rd Session of the FAO Conference in 1985. The Code was amended in 1989 at the 26th Session of the FAO Conference to incorporate the principle of *Prior Informed Consent.*

It was recognized that in order to remain relevant the Code must evolve in order to reflect changing needs of countries and that there was a need to monitor progress in the observance of the Code.

One of the basic functions of the Code is to serve as a point of reference, particularly until such time as countries have established adequate regulatory infrastructures for pesticides. The objectives of the Code are to set forth responsibilities and establish voluntary standards of conduct for all public and private entities engaged in or affecting the distribution and use of pesticides. The Code suggests how to distribute the responsibilities between government, industry and others.

The twelve articles of the Code are supported by a set of detailed technical Guidelines which provide guidance on their implementation. The Guidelines include such topics as: registration and control of pesticides; legislation; personal protection for those working with pesticides; aerial spraying; disposal of waste pesticides and containers; tender procedures for the procurement of pesticides; etc. These documents have been distributed widely and are frequently the basis for the development of policies and practices in countries looking to initiate or improve a pesticide management scheme.

Implementation of the Code—Establishing a Regulatory Framework at the National Level. The Code of Conduct and its supporting technical guidelines provide a framework for the management of pesticides within a crop protection programme. In order that countries are in a position to implement specific provisions of the Code, FAO provides guidance and basic technical assistance in establishing and strengthening regulatory programmes and the facilities required. This has taken the form of regional projects in Asia and the Pacific and the Caribbean (1988–93, funded by Japan) and Africa (preliminary phase completed in 1993; a final draft of the major phase is in preparation). A project for Central America is planned for 1995/96. The objectives of the Regional projects include:

- to assist governments in establishing simple but effective national pesticide registration and control schemes adaptable to their social, political, environmental and economic needs;
- to assist governments in strengthening their ability to enforce national pesticide regulatory schemes more effectively;
- to promote harmonization of pesticide registration requirements and the exchange of information on pesticide regulatory issues and other technical matters of importance at the regional level;
- to promote the safe and efficient use of pesticides through national training programmes for farmers, extension workers, retailers and medical personnel.

Since 1990 there have been 24 Technical Assistance Projects to assist countries implement aspects of the Code, ranging from projects at the national level to strengthen national registration systems to sub-regional projects on the harmonization of requirements and procedures for the registration of pesticides.

Two questionnaires have been sent to countries to monitor progress on the application of the Code, one in 1986 and a second in 1994. Results from these surveys have helped to identify those areas where technical assistance needs are greatest and follow-up activities will have most impact on reducing risks.

Efficacy test protocols. An integral part of the implementation of the Code in some countries has been the development of efficacy test protocols as part of a programme to promote harmonization of pesticide data requirements within a region. The harmonization of data requirements for registration among countries in a region makes more efficient use of the limited resources and contributes to better pesticide management.

Specifications for plant protection products. An ongoing activity aimed at ensuring the quality of pesticides relates to FAO Specifications for Plant Protection Products. FAO has published technical specifications for more than 200 pesticides and related formulations. They provide: a basic standard of quality for the buying and selling of pesticides: assistance in the official approval and acceptance of pesticides; protect responsible vendors against inferior products; and a link between biological efficacy and specification requirements. A *Manual on the Development and Use of FAO Specifications for Plant Protection Products* contains detailed definitions and other background information to facilitate the development and use of the specifications.

Pesticide disposal. The problem of pesticide disposal exists at the farm level and on a large scale in many developing countries where considerable stocks of obsolete and outdated pesticides have accumulated. These stockpiles have accumulated for a variety of reasons and represent a potentially significant hazard to human health and the environment. A three-year project in Africa and the Near East was initiated in 1994. Activities include a survey of the stockpiles of obsolete or unwanted pesticides and the development of a pilot project in selected countries to determine the best methods for disposal. A series of technical guidelines is under development in cooperation with UNEP and WHO relating to the disposal of pesticide containers and pesticide waste as well as how to more effectively manage pesticide stocks in order to prevent the accumulation of obsolete stocks in the future.

Prior Informed Consent Procedure (PIC). The principle of Prior Informed Consent (or PIC) was incorporated into the International Code of Conduct on the Distribution and Use of Pesticides and the London Guidelines for the Exchange of Information on Chemicals in International Trade in 1989. The FAO/UNEP Joint Programme on the Implementation of the PIC Procedure was established in 1990.

PIC is not a blacklist of compounds. Its objective is to provide countries with detailed information on the chemical in question and on reasons for bans or severe restrictions in the form of Decision Guidance Documents (DGDs). In the light of this information, countries are asked to assess their position and the risks associated with the pesticide, and to make timely and informed decisions as to the importation and use of the chemical concerned. The import decisions are transmitted to all other participating countries every six months.

Integrated Pest management (IPM). The plan of action adopted at the United Nations Conference on the Environment and Development (UNCED) includes a Chapter on Sustainable Development (Chapter 14) within which the importance of Integrated Pest Management (IPM) is emphasized.

Integrated pest management maximizes the value of the crop to farmers while minimizing environmental and health risks in a sustainable way. Farmers practising IPM rely on a range of methods including biological control, host plant resistance, chemical pesticides and cultural practices. It is not a technology but a problem-solving process whereby farmers learn from their own experiences and carry out their own 'research' to create solutions to their crop management problems.

The FAO Intercountry Programme for the Development and Application of IPM in Rice in South and South East Asia has been in operation for twelve years. Most of the eleven participating countries (Bangladesh, Cambodia, China, India, Indonesia, Laos, Malaysia, Philippines, Sri Lanka, Thailand and Vietnam) have established national programmes in IPM. There are several examples where the implementation of IPM has resulted in a decrease in the use of insecticides by up to 50%, with no decrease in yield.

The IPM strategy and training methodologies developed in Asia are being extended to other crops and to countries in other regions. Regional projects in Asia (for cotton and vegetables) and Africa (for vegetables), as well as at the national level in Ghana, Cote d'Ivoire and Burkina Faso (rice), Zimbabwe (cotton), Sudan (wheat and vegetables), Bangladesh, India and the Republic of Korea have been initiated.

Efforts are being made to coordinate activities and resources for IPM development and implementation among various agencies. A joint programme has been established between FAO and the World Bank. Negotiations are ongoing to include sponsorship by UNEP and UNDP. This 'IPM facility' will draw upon local, national and international expertise, knowledge and resources to facilitate the process of identification, design and implementation of national IPM programmes. Such a mechanism will provide more rapid and effective response to the needs of farmers and national IPM programmes and greater efficiency of resource use by national governments and development agencies.

International Labour Organisation (ILO)

International Labour Office
4 route des Morillons
CH-1211 GENEVA 22
Tel: (22) 799 61 11
Telex: 415 647 ILO CH
Fax: (22) 798 86 85

Aims/Activities

The International Labour Organization (ILO) was established in 1919 and became the first UN Specialized Agency in 1946. Its mandate is to raise standards of living worldwide, to promote social justice and reform, and to seek adequate protection for the life and health of workers in all occupations. ILO works towards these goals by setting international labour standards, by promoting research, and by providing technical assistance to Member countries. The organization has a tripartite structure in which representatives of labour, employers as well as government participate in the decision-making process at all levels. As of February 1994. ILO had a membership of 170 States. The Headquarters of ILO is located in Geneva, Switzerland.

Most of ILO's activities related to chemical safety are conducted through the Working Conditions and Environment Department. The Working Conditions and Environment Department is concerned with working conditions, the working environment and well-being of workers. Activities are guided by the principle that work should take place in a safe and healthy environment; conditions of work should be consistent with workers' well-being and human dignity; and work should offer real possibilities for personal achievement, self-fulfilment and service to society. Activities conducted within the Department include setting and implementing international labour standards, collecting and disseminating information, and providing technical assistance to member States.

International Occupational Safety and Health Information Service (CIS)

International Labour Office
CIS J. Takala
4 route des Morillons
CH-1211 GENEVA 22
Switzerland
Tel: (41) 22 7996740
Telex: 415647 ILO CH
Fax: (41) 22 7986253

Aims/Activities

The International Occupational Safety and Health Information Centre is a Branch of the Working Conditions and Environment Department. It is concerned with the capacity of member states to apply new information in the formulation and implementation of occupational safety and health policies and standards. The exchange of safety and health information is organized and executed through the activities carried out under the Centre in Geneva with the assistance of designated institutions (CIS national and collaborating centres) in 85 member states. CIS collects, processes and publishes abstracts of all important safety and health literature, including documents on chemical safety, which it receives through the network of collaborating institutions. New toxicity information is a current area of priority interest. Training and technical assistance activities promote the implementation of ILO standards, the exchange of information and the provision of technical advisory services in safety and health legislation, chemical safety, and worksite safety measures. These activities of the Centre are organized in close coordination with the Occupational Safety and Health Branch.

CIS, established in 1959, is a branch of the Working Conditions and Environment Department of the International Labour Office. CIS, assisted by over 70 affiliated National Centers, collects and scans the world literature on occupational safety and health and disseminates this information at an international level. Abstracts of the literature are prepared and stored in the CIS bilingual (English and French) database (CISDOC or CISILO), which contains more than 40,000 records and is updated with more than 2000 records every year. Information from this database is available in hard-copy, for online access and on CD-ROM. CIS disseminates information on chemicals and pesticides through chemical safety information sheets.

ILO Publications

- Major Hazard Control, A practical manual. 2nd ed. Geneva, 1990. (ILO Publications) ISBN 92–2–106432–8.
- Safety and Health in the Use of Agrochemicals: A Guide. 1st ed. Geneva, 1991. (ILO Publications) ISBN 92–2–107281–9.
- Guide on Training of Trainers in Labour Inspection. Geneva, 1990. (ILO Publications)
- Guidelines for the Establishment of National CIS Centres. (ILO–CIS).
- Accident Prevention. 2nd ed. Geneva, 1986. (ILO Publications) ISBN 92–2–103392–9.

Codes of Practice:

- Safety and Health in the Use of Chemicals at Work. A Code of Practice. 1st ed. Geneva, 1993. (ILO Publications)
- Occupational Exposure Limits for Airborne Toxic Substances, No. 37. 3rd Edition. Geneva, 1991. (ILO Publications)

- Classification, Identification and Labelling of Chemicals. Geneva, 1992. (ILO–CIS).

Newsletters and Brochures:

- CIS Bulletin, Safety and Health at Work.
- East African Newsletter on Occupational Health and Safety. Helsinki, Finland, 1987–1990. (ILO and Finnish Institute of Occupational Health).
- African Newsletter on Occupational Health and Safety. Helsinki, Finland, 1991–Ongoing. (ILO and Finnish Institute of Occupational Health).
- Asian Newsletter on Occupational Safety and Health. Helsinki, Finland, 1992–Ongoing. (ILO and Finnish Institute of Occupational Health).

Databases:

- The CIS Database.
- Worksite Safety Information Sheets.
- Chemical Safety Data Sheets.
- Listing of Training Materials and Training Institutions.
- Chemdata Computer Databases. (Information on major hazard chemicals and exposure limits).
- Country Information Databases. (Pilot software produced for the African Training and Information Project on Occupational Safety and Health.)

Other Support Services/Materials:

- Basic Safety Training Package on Occupational Safety and Health.
  Questionnaires for pre-course evaluation, course manual, AV-materials, group work instructions, reference textbooks, student materials and course evaluation materials.
- Basic Training Package on Occupational Hygiene.
  Questionnaires for pre-course evaluation, course manual, AV-materials, group work instructions, reference textbooks, student materials and course evaluation materials.
- Package on Training of Trainers.
  Questionnaires for pre-course evaluation, course manual, AV-materials, group work instructions, reference textbooks, student materials and course evaluation materials.

ILO–CIS Bulletin–Safety and Health at Work (6 times annually in French and English)

United Nations Institute for Training and Research (UNITAR)

UNITAR
Palais des Nations
1211 Geneva
Switzerland
Tel: 41 22 798 8400
Fax: 41 22 733 1383

Program Information

The United Nations Institute for Training and Research (UNITAR), jointly with the FAO/UNEP Programme on Prior Informed Consent, provides a variety of training and technical support services to Designated National Authorities and other concerned governmental and non-governmental bodies and organizations concerned with the implementation of the PIC procedure in developing countries. Workshops are organized at the regional, sub-regional and national levels and specialized training materials and manuals have been developed. A specific training manual titled 'Strengthening Capabilities in Countries to Make PIC Import and Related Chemicals Risk Management Decisions' is available as a draft and is currently under review. In addition, UNITAR has recently initiated two programmes which are closely linked to the international and national processes to implement Chapter 19 of Agenda 21.

1. *The UNITAR Programme to Assist Countries to Prepare National Profiles to Assess the National Infrastructure for the Sound Management of Chemicals* provides guidance to countries to document and identify strengths and weaknesses of their national chemical management schemes by preparing a comprehensive National Profile through a participatory process. This Profile may in turn serve as a basis for developing and implementing national action programmes for the sound management of chemicals, including pesticides. The programme is implemented under the umbrella of the Inter-Organization Programme for the Sound Management of Chemicals, a collaborative programme of FAO, OECD, ILO, UNEP, UNIDO and WHO, and it is linked to the work of the Inter-governmental Forum on Chemical Safety. A guidance document has been prepared and pilot studies have been initiated in the Czech Republic, Mexico, Guinea, and Zambia.

2. The *UNITAR Programme to Facilitate the Establishment of Pollutant Release and Transfer Registers in Developing and Industrializing Countries* promotes pollution prevention through establishing national systems to register pollutant releases and transfers, which may include pesticides emissions and transfers from point and non-point sources. Working closely with the Pollution Prevention and Control Group of OECD and several international organizations, UNITAR has initiated a series of pilot studies in Egypt, the Czech Republic and Mexico to assess the feasibility of introducing PRTRs as an innovative environmental management tool in industrializing and developing countries.

United Nations Environment Programme (UNEP)

United Nations Environment Programme
P.O. Box 30552
NAIROBI
Kenya
Tel: (2) 333930
Telex: 22068

Industry and Environment Office
United Nations Environment Programme
17 rue Margueritte
F-75017 PARIS
France
Tel: (1) 47661640
Telex: 650273

Liaison Office:

United Nations Environment Programme
ALCOA Building, Room A-3630
866 United Nations Plaza
NEW YORK NY 10163
USA
Tel: (212) 754-8139
Telex: 420544

International Register of Potentially Toxic Chemicals
15 Chemin des Anemones
Case postale 356
CH-1219 Geneva
SWITZERLAND
Tel: (41) 22 979 9111
Telex: 415 465 UNE CH
Fax: (41) 22 797 34 60

Aims/Activities

The United Nations Environment Programme was established in 1972 following a recommendation of the United Nations Conference on the Human Environment, held in Stockholm in 1972. As mandated by the 1972 Stockholm Conference, UNEP catalyses and coordinates environmental work within and outside of the U.N. System. UNEP also provides leadership and encourages partnership in caring for the environment by inspiring, informing and enabling nations and people to improve their quality of life without compromising that of future generations.

International Register of Potentially Toxic Chemicals [UNEP/ Chemicals (IRPTC)]

The International Register of Potentially Toxic Chemicals (IRPTC) operates several activities and services related to chemicals management. These include the implementation of the London Guidelines for the Exchange of Information on Chemicals in International Trade, the operation of the IRPTC Databank and a global network for the exchange of information through network partners, and the implementation of a Toxic Chemicals and Waste Management Capacity Building Programme.

Implementation of the London Guidelines. IRPTC has the responsibility to facilitate the implementation of the London Guidelines for the Exchange of Information on Chemicals in International Trade, in particular, the Prior Informed Consent (PIC) procedure. Jointly with FAO, IRPTC operates the FAO/UNEP Joint Programme on the Operation of PIC. Under this programme Decision Guidance Documents for chemicals included in the PIC procedure are issued and a databank with all PIC-relevant information (e.g. import decisions) is maintained.

IRPTC Databank and Its Use in Decision Making. Data on the hazards of chemicals and the methodologies to assess such hazards are the cornerstone of chemicals management. The IRPTC databank, which is available to users as a personal computer version, contains *Data Profiles* on over 800 chemicals. Information contained in the files is meant to provide a basis for sound risk management decisions, in particular, for chemicals subject to the PIC procedure. The *Legal File* contains national and international regulatory and legal information on over 8000 chemicals. Together with its partners, IRPTC is developing methodology for the use of these data in decision making e.g. for priority setting and hazard assessment.

IRPTC's Global Network of Partners. IRPTC has established several global networks which address specific issues of international information exchange: National Correspondents assist IRPTC in information gathering and dissemination: Contributing Network Partners assist with the collection and validation of data and provide scientific and technical expertise upon request: and Designated National Authorities are governments institutions which are responsible at the national level for implementing the London Guidelines.

IRPTC's emerging Global Network of Partners for Electronic Exchange of existing information is one of the keys for future information exchange and data management. Presently, IRPTC is engaged with UNECE, IPCS and specific member countries to identify important data sources and to explore the possibility of standardizing data exchange mechanisms within the framework of UN Electronic Data Interchange (EDI) programme.

Toxic Chemicals and Waste Management Capacity Building Programme. In 1991 IRPTC started a process of consolidating and strengthening its training and technical assistance activities in countries of Central and Eastern Europe and in developing countries. Building on its earlier capacity-building activities in Central and Eastern Europe (see Section II, Entry #s: A3 and A5), IRPTC, in light of political changes resulting in the establishment of many new states and bringing a larger number of countries in economic transition, started to develop a new concept and new approaches to help improve chemicals management schemes for these countries. In developing countries, agreement on co-operation with UNITAR and the initiation of a joint UNEP/UNITAR Training Programme on the Implementation of the London Guidelines in developing countries constituted a major step in strengthening IRPTC capacity building activities in the developing regions.

Publications

Manuals, Guides and Technical Publications:

- IRPTC Personal Computer Databank, Version 2.0, User's Manual. Geneva, 1994. (UNEP?IRPTC)
- IRPTC Personal Computer Databank, Version 2.0, Administrator's Manual. Geneva, 1994. (UNEP/IRPTC)
- IRPTC Legal File 1994/1995. Geneva.
- Guidance for Governments. (Joint UNEP/FAO Programme for the Operation of Prior Informed Consent)
- Decision Guidance Documents, Issues No. 1–3. (Joint UNEP/FAO Programme for the Operation of Prior Informed Consent)
- Survey of Technical Guidance Material Developed by International Organizations, OECD and EU for Priority Setting of Existing Chemicals.
- International Survey of Classification Systems for Chemicals Based on Their Toxicity and Hazards. (Prepared for Printing)
- IRPTC Waste Management—Technical Manual. (In Preparation)
- Training Material for Priority Setting of Existing Chemicals for Developing Countries. (In Preparation)

Newsletters and Brochures:

- IRPTC Bulletin. (English, French, Russian and Spanish)
- International Register of Potentially Toxic Chemicals (IRPTC). 1990. Brochure. (English, French, Russian and Spanish)

Databases:

- IRPTC Personal Computer Databank Version 2.0. Geneva, 1994. (UNEP/IRPTC)

United Nations Industrial Development Organisation (UNIDO)

Vienna International Centre
P.O. Box 300
A-1400 VIENNA
Austria
Tel: (1) 211310
Telex: 135612
Fax: (1) 232156/2140414

Aims/Activities

UNIDO was established in 1966 as an autonomous organ within the UN Secretariat. It was given the status of UN Specialized Agency in 1986. UNIDO's mandate is to promote industrial development in developing countries by means of technical assistance, training, the exchange of information, the promotion of investment, national and regional planning and the transfer of technology. UNIDO conducts a variety of technical assistance and field projects aimed at developing, expanding and modernizating specific industries. In addition, UNIDO provides advice to Governments on economic planning and assists them in obtaining financing for specific industrial projects. UNIDO coordinates all activities of the UN system relating to industrial development. In 1985 the UN General Assembly adopted the Lima Declaration and Plan of Action for UNIDO requesting the organization to mobilize human and material resources to cope with problems which threaten the environment and giving it the additional mandate to investigate and promote the choice and application of environmentally sound technology.

UNIDO has a major programme in risk reduction in pesticide development in developing countries. Regional networking in Asia has been successfully accomplished through a network called *Regional Network on Pesticides for Asia and the Pacific (RENPAP)*. The network comprising fifteen member countries in Asia covers a wide spectrum of areas to minimize risks associated with pesticides. Six technical coordinator units have been created and their capacity strengthened to assist the member countries. The topics covered are:

- User and environmentally friendly pesticide formulations and quality control—India
- Impurities in Technical Materials—South Korea
- Bio- and Botanical Pesticides—Thailand
- Application Technology—Malaysia
- Industrial Safety/Waste Management—Indonesia
- Industrial Hygiene and Occupational Health—Philippines
- Ecotoxicology—Pakistan
- Data Collection and Dissemination—India/Thailand

The regional project is also linked to national projects in India, China, and Pakistan on pesticides. The network has created a greater awareness of safety and ecotoxicology and has been a catalyst in causing countries such as India and the Philippines to place bans or restrictions on many persistent and toxic pesticides. A similar network in Afro-Arab countries is being set up.

On a global basis, UNIDO has developed integrated international safety guidelines for pesticide formulation in developing countries, which cover safety, health and environmental aspects of pesticide formulation and implementation on a country basis.

UNIDO co-ordinates all UN activities relating to industrial development in developing countries, including pesticide plants, especially formulation plants. Technical assistance and advice are given on feasibility studies, technology assessment, transfer of appropriate technology, plant design, quality control, integrated safety and water management. Emphasis is placed on regional networking on safety in pesticide production/use, safety guidelines and effluent limitation guidelines. UNIDO also promotes bio/botanical, user/environment safe pesticides and formulations, and ecotoxicology studies.

Publications

- Recent Development in the Field of Pesticides and their Application to Pest Control (Proceedings of the International Seminar held at Shenyang. China, 8–12 October 1990)
- International Integrated Safety Guidelines for Pesticide Formulation in Developing Countries (free)
- Proceedings of Expert Group Meeting to Develop Integrated International Safety Guidelines for Pesticide Formulation in Developing Countries held in Brussels, 24 February–6 March, 1992.
- Technical Report of Workshop on Bio/Botanical Pesticide Development held in Bangkok, 11–18 December 1991
- Pesticide Warning Card (free)
- Regional Network on Pesticides for Asia and the Pacific (RENPAP Gazette)

World Health Organisation (WHO)

Avenue Appia 20
CH-1211 GENEVA 27
Switzerland
Tel: (22) 7912111
Telex: 415416 OMS
Fax: (22) 7910746

Aims/Activities

The World Health Organization was founded in 1948 as a Specialized Agency of the U.N. WHO's general objective is the attainment of the highest level of health for all people. It works towards this goal through a broad range of operational programmes and by addressing Member States through recommendations and regulations. As of August 1993, WHO had 187 Member Countries and 2 Associate Members. The Headquarters of WHO is located in Geneva, Switzerland.

The 'WHO Global Strategy for Health and the Environment', endorsed by the World Health Assembly in May 1993, provides a unifying framework for all WHO substantive and technical assistance programmes in the field of health-and-environment, including chemical safety. It defines Organization-wide objectives and action programmes and provides the basis for WHO programmes at Headquarters, in the regions and in countries to devise workplans in order to implement the health-and-environment activities of Agenda 21. The major components of the Strategy include: an enhanced programme for the promotion of environmental health; an expanded Programme for the promotion of chemical safety; and broader action and collaboration on matters affecting health and environment throughout WHO. All WHO health-and-environment programmes are coordinated by the Executive Director, Health and Environment. Relevant activities are carried out through the Programme for the Promotion of Chemical Safety, the Office of Global and Integrated Environmental Health, and the Office of Operational Support in Environmental Health.

The Division of Health Protection and Promotion is concerned with the health of a number of specific population groups and health problems, such as the health of the elderly and workers', and prevention of injuries, blindness and deafness. Relevant activities of the division are carried out by the Office of Occupational Health.

The Division of Food and Nutrition is mainly concerned with chemical safety as it relates to food safety. Relevant activities of the division are carried out by the Food Safety Unit.

Office of Occupational Health

The Office of Occupational Health is part of the Division of Health Protection and Promotion. The Office implements the WHO Workers' Health Programme (WHP) which has the objectives of controlling occupational health risks and protecting and promoting the health of the working populations as well as the humanization of work. One of the main focuses of WHP has been to develop and publish guidelines dealing primarily with the monitoring of workers' health and environment. Presently, efforts are being made to finalize guidelines for: *Biological Monitoring of Chemical Exposure at the Workplace*; *Occupational Respiratory Diseases Induced by Organic Dust and Sensitizing Agents—Health Screening and Surveillance*; and *Reference Method and Quality Assurance for Evaluation of Fibres in Workplace Air*. Additional guidance materials relating to chemical safety to be developed include: the updating of the WHO *Standard protocol in Field Studies of Exposure to Pesticides*; *Biological Monitoring of Occupational Exposure to*

Carcinogenic/Mutagenic Chemicals; and *Guiding Principles and Guidance Values for Occupational Exposure Limits for Chemicals*. The Office of Occupational Health collaborates with the ILO in the *Joint ILO/WHO Committee on Occupational Health*. The Committee helps to ensure that the occupational health programmes of each organization are mutually supportive and not duplicative.

The Food Safety Unit

Within the Division of Food and Nutrition, the Food Safety Unit seeks to reduce morbidity and mortality due to foodborne disease through the promotion of food that is safe, sound, wholesome and fit for human consumption. To this end, the Food Safety Unit supports national food safety programmes, including policy development, capacity building, awareness promotion, and information gathering and transfer with a view to protecting human health against hazards associated with biological and chemical contaminants and additives in food. Within WHO, the Food Safety Unit is responsible for exposure assessment, risk characterization and risk management aspects of chemicals in food and for the operative recommendations of UN Conference on Environment and Development and the International Conference on Nutrition which are pertinent to food safety. Through the six WHO Regional Offices, regional and national training courses and seminars are regularly convened on a variety of food safety topics, including food law and regulations, inspection methods and laboratory analysis of contaminants. In addition, consultant services, fellowships and equipment and supplies are provided to support the development of national food safety programmes.

Manuals, Guides and Technical Publications:

- WHO/FAO Data Sheets on Pesticides. (FAO and WHO)
- Principles for the Toxicological Assessment of Pesticide Residues in Food. Environmental Health Criteria, No. 104. 1990.
- Guidelines for Predicting Dietary Intake of Pesticide Residues. 1989
- Guidelines for the Study of Dietary Intakes of Chemical Contaminants, 1985. (WHO Offset Publication, No. 87).

International Agency for Research on Cancer

The International Agency for Research on Cancer (IARC) was established by WHO in 1965 and has its Headquarters in Lyon, France. IARC conducts and directs research designed to identify causes of human cancer and define the best lines for preventative action. Research conducted at IARC includes both epidemiological studies on environmental factors that may cause cancer and laboratory studies that promote better knowledge of the disease process at the molecular and biochemical levels. IARC also participates in the IPCS in the field of chemical carcinogenicity. IARC publishes a series of

Monographs on the Evaluation of Carcinogenic Risks to Humans which may assist national and international authorities in making risk assessments and in formulating decisions regarding preventive measures. IARC also publishes volumes containing summaries of worldwide cancer incidence and mortality data, and results of analytical research.

IARC Programme for Education and Training. The IARC Programme for Education and Training conducts courses for cancer research personnel and public health workers in the fields of cancer epidemiology, statistical methods in epidemiology, carcinogenicity testing, molecular biology for exposure assessment and early effects of carcinogenesis, genetic toxicology and safe laboratory practices. The training courses are often conducted jointly with other international organizations or with collaborating institutions. In addition, IARC awards research training fellowships in research fields related to the Agency's own programme.

UNITED NATIONS INTER-AGENCY PROGRAMMES

International Programme on Chemical Safety (IPCS)

Avenue Appia 20
CH-1211 GENEVA 27
Switzerland
Tel: (22) 7912111
Telex: 415416 OMS
Fax: (22) 7910746

Aims and Activities

The International Programme of Chemical Safety (IPCS) is a joint programme of ILO, WHO, and UNEP. In this capacity IPCS provides coordinating services for the various activities of ILO, UNEP and WHO conducted under the umbrella of IPCS. In addition, IPCS performs technical work, on behalf of the three organizations, which contributes directly to the implementation of the IPCS objectives. The activities of IPCS are carried out through two units: the Unit for Prevention and Treatment (PPT), and the Unit for Assessment of Risks and Methodologies (ARM).

Risk Evaluation of Priority Chemicals. Evaluation of the risks to human health and the environment posed by specific chemicals is one of the main objectives of the IPCS. Results of these evaluations are disseminated in the form of Environmental Health Criteria Documents and Health and Safety Guides. Chemicals are chosen for evaluation on the basis of adverse effects to human health or the environment, degree of exposure to humans or the environment, size and nature of the population at risk, and international concern. Other outputs of the risk evaluations are: the International Chemical Safety Cards (prepared jointly with CEU), promotion of safe use of chemicals in the workplace; Poisons Information Monographs (PIMs) to aid the medical profession in diagnosis and treatment of poisoning; and Monographs on Toxins of Plant Origin and Pesticide Data Sheets, to provide guidance on safe use of pesticides. Training activities and workshops are organized on the use of these evaluations in chemical risk management.

JECFA Evaluations of Food Additives. The joint FAO/WHO Expert Committee on Food Additives (JECFA) provides advice to Member States and the Codex Alimentarius Commission on the level at which specific chemicals, principally food additives, can be safely tolerated in the human body. IPCS-PCS is responsible for the technical work, while the WHO Food Safety Unit is responsible for the subsequent work and recommendations of the Codex Alimentarius Commission. JECFA publishes specifications for the identity and purity of food additives and toxicological monographs which summarize safety data on selected food additives.

Development of Health Risk Assessment Methodologies. This programme area has the objective to facilitate the development and validation of internationally accepted approaches and methods for testing, assessing and predicting the effects of chemicals on human health and the environment. Technical reports and monographs published through the Environmental Health Criteria series cover a wide variety of methodological issues. Training activities and seminars focus on chemical risk assessment processes and procedures and address in particular the needs of toxicologists and environmental health professionals in developing countries.

Chemical Poisoning Prevention and Treatment. This programme area aims at strengthening national capabilities to respond to emergencies and accidents involving chemicals. Activities include, *inter alia*, the preparation of guidance material for poisons control and its use in establishing poisons control facilities in countries: the evaluation and harmonization of information needed for the diagnosis and treatment of poisonings; the development of a compatible, computerized information system (INTOX); and the development and training of manpower needed to operate poison control centres and related facilities.

The programme maintains an informal network of toxicologists, health care professionals and poison control centres worldwide. The technical assistance and training activities provide advice to health care professionals and staff of poison control centres on the prevention and treatment of chemical poisonings and on the use of the INTOX software.

Health Aspects of Chemical Emergencies. This programme area has the objective to provide guidance and assistance to member states on the health aspects of chemical accidents. The primary responsibility of the IPCS-PCS in this area is to develop guidance documents on the health aspects of chemical accidents. Activities within the programme are integrated with those of the Prevention and Treatment of Poisonings Programme. Strengthening the role of poison centres in responding to major chemical emergencies and contingency planning is also an important element of this programme. Training activities on health aspects of chemical accidents are often conducted alongside the training activities of the Prevention and Treatment of Chemical Poisonings Programme.

Publications

Manuals, Guides and Technical Publications:

- Guidelines for Poisons Information Centres: Their Roles in Poisons Control Programmes. Geneva, 1994.
- Manual on Simple Analytical Toxicology Techniques. Geneva, 1994.
- IPCS/INTOX Poison Information Package. (WHO/PCS, Canadian Centre for Occupational Health and Safety (CCOHS) and the Centre de Toxicolgie due Québec (CTQ))
- Health Aspects of Chemicals: Guidance on Chemical Accident Awareness, Preparedness and Response for Health Professionals and Emergency Responders. (IPCS/OECD/UNEP IE-PAC/WHO-ECEH)
- Chemical Safety Matters. (IUPAC/PICS joint publication by CUP)

Newsletters and Brochures:

- IPCS News.
- The International Programme on Chemical Safety (IPCS). Geneva. (English, French, Spanish, German, Russian and Chinese)

Available from: World Health Organisation Publications, Distribution and Sales, 1211 Geneva, Switzerland (who can also supply a full list of the pesticides covered in each series, and their prices, which include a 30% reduction for developing countries).

Codex Committee on Pesticide Residues (CCPR)

Secretariat
Joint FAO/WHO Food Standards Programme
FAO
Via delle Terme di Caracalla
I-00100 ROME
Italy
Tel: (6) 57971
Telex: 610181 FAO, 611127 FAO
Fax: (6) 5146172: (6) 6799563

Aims/Activities

The objectives of the CCPR of the Codex Alimentarius Commission (CAC) are:

1. To establish maximum limits for pesticide residues in specific food items or in groups of foods.
2. To establish maximum limits for pesticide residues in certain animal feed moving in international trade where this is justified for reasons of protection of human health.
3. To prepare priority lists of pesticides for evaluation by the Joint FAO/WHO Meeting on Pesticide Residues (JMPR).
4. To consider methods of sampling and analysis for the determination of pesticide residues in food and feed.
5. To consider other matters in relation to the safety of food and feed containing pesticide residues.
6. To establish maximum limits for environmental and industrial contaminants showing chemical or other similarity to pesticides, in specific food items or groups of foods.

International Environmental Information System (INFOTERRA)

INFOTERRA Programme Activity Centre
UNEP
P.O. Box 30552
NAIROBI
Kenya
Tel: (2) 333930
Telex: 22068 UNEP

Aims/Activities

INFOTERRA is an environmental information network established by the United Nations Environment Programme (UNEP), co-ordinated through national focal points in 123 countries.

Joint FAO/WHO/UNEP Food Contamination Monitoring Programme (GEMS/FOOD)

WHO
Avenue Appia 20
CH-1211 GENEVA 27
Switzerland
Tel: (22) 7912111
Telex: 415416 OMS
Fax: (22) 7910746

Aims/Activities

Since 1976, the Joint UNEP/FAO/WHO Food Contamination Monitoring and Assessment Programme (commonly referred to as GEMS/Food) has collaborated with member states in the collection and assessment of data on levels of certain pesticides, heavy metals and industrial and naturally occurring toxicants in foods and the total diet in order to provide information on human exposure to potential hazardous chemicals in the environment. Training and technical assistance can be provided through the network of GEMS/Food Collaborating Centres and participating institutions which are located in 59 countries world-wide.

The Collaborating Centres are requested to provide data on chlorinated pesticides (DDT-complex, HCH-isomers, heptachlor and its epoxide, aldrin and dieldrin, HCB), PCBs, organophosphorus pesticides (malathion, parathion, methyl parathion, diazinon), lead, cadmium and aflatoxins in a variety of foods in the diet. Summaries of the monitoring data received on contaminants in individual foods are issued by country, by contaminant and by food groups for data received. Monitoring data on contaminants in the daily diet of various countries have also been issued.

Analytical quality assurance studies are being carried out to promote comparability and quality control of the data submitted under the Joint Programme. Such inter-laboratory quality assurance studies are an integral and essential part of this monitoring programme.

Guidelines for the Study of Dietary Intakes of Chemical Contaminants have been issued and made available to the Joint FAO/WHO Collaborating Centres for Food Contamination Monitoring as well as to Codex Contact Points and others to encourage collection of data on the intake of chemical contaminants from food with a view to evaluating the potential risk to human health from such exposure.

Joint FAO/WHO Food Standards Programme, Codex Alimentarius Commission (CAC)

FAO
Via delle Terme di Caracalla
I-00100 ROME
Italy
Tel: (6) 57971
Telex: 610181 FAO

Aims/Activities

At the global level, WHO collaborates with FAO in the inter-governmental Codex Alimentarius Commission which provides international risk management recommendations related to food additives, pesticide and veterinary drug residues and other food contaminants based on scientific evaluations of

JECFA and JMPR. As a consequence of the GATT-Uruguay Round agreement, Codex recommendations, guidelines and codes of practice will be used as the basis for arbitration involving safety and health requirements employed as non-tariff barriers to trade. In addition, WHO convenes expert consultations on new and emerging food hazards to provide Member States with guidance in risk perception, assessment and management. In particular, a series of training courses have been organized to promote the application of the Hazard Analysis Critical Control Point system in the analysis and management of food safety hazards at all stages of the food chain.

CAC is involved in the development of standards for food products which are important in international trade, general food standards, and recommendation for food handling and production. The organisation has six general committees (General principles, Food labelling, Food additives, Pesticide residues, Food hygiene and Methods of analysis and sampling), four regional co-ordinating committees (for Africa, Asia, Europe, Latin America and the Caribbean) and 17 commodity committees for special types of food (Milk and milk products; Fats and oils; Edible ices; Processed fruits and vegetables; Fruit juices; Quickfrozen foods; Fish and fishery products; Foods for special dietary uses; Meat; Processed meat and poultry products; Meat hygiene; Cereals, pulses and legumes; Vegetable proteins; Cocoa products and chocolate; Soups and broths; Sugars; Natural mineral waters).

Codex Committee on Pesticide Residues (CCPR). The maximum residue levels proposed by the JMPR are considered by the *Codex Committee on Pesticide Residues (CCPR)*. The CCPR, a subsidiary body of the *Codex Alimentarius Commission (CAC)*, is an intergovernmental meeting whose prime objective is to reach agreement between governments on maximum residue limits (MRLs) for pesticide residues in food and feed commodities moving in international trade. The MRL proposals are considered by the CCPR as part of eight-step procedure which provides opportunity for discussion and comment by national governments and other interested organizations. The CCPR recommends MRLs to the biennial meeting of the CAC, for adoption as Codex maximum residue limits (CXLs).

As the Codex Alimentarius Commission, whose primary purpose is to protect the health of consumers and to ensure fair practices in the food trade, is recognized in the World Trade Organization's Agreement of the Application of Sanitary and Phytosanitary Measures as the only body establishing international food standards, Codex CXLs will be used as references in cases of trade disputes between countries regarding pesticide residues in food moving in international trade.

The parent organizations of the Codex Alimentarius Commission, FAO and WHO, have recently convened expert consultations on the application of risk analysis to food standards issues and on predicting dietary intake of pesticide residues.

Joint Meetings of the FAO Panel of Experts on Pesticide Residues and the Environment and the WHO Expert Group on Pesticide Residues (JMPR)

WHO Secretariat	International Programme on Chemical Safety Division of Environmental Health World Health Organisation CH-1211 GENEVA 27 Switzerland Tel: (22) 7912111 Telex: 415416 OMS Fax: (22) 7910746
FAO Secretariat	Plant Protection Service Food and Agriculture Organisation of the United Nations Via delle Terme di Caracalla I-00100 ROME Italy Tel: (6) 57971 Telex: 610181 FAO, 611127 FAO Fax: (6) 5146172

Aims/Activities

The Joint FAO/WHO Meetings on Pesticide Residues (JMPR) consist of two separate bodies: The WHO Expert Group on Pesticide Residues and the FAO Panel of Experts on Pesticide Residues in Food and the Environment.

JMPR conducts studies to evaluate the consumer hazard arising from the use of pesticides. This involves examination of the toxicological and other pertinent data on those pesticides known to leave residues in food, even when used in accordance with good agricultural practice. JMPR provides advice to Member states and to the Codex Alimentarius Commission on the safe level and effects of pesticides in the body. It has met annually since 1963 to conduct scientific evaluations of pesticide residues in food. It provides advice on the acceptable levels of pesticide residues in food moving in international trade. The JMPR consists of experts drawn from governments and academic circles. They attend as independent internationally-recognized specialists who act in a personal capacity and not as representatives of national governments.

The FAO segment of the JMPR, the *FAO Panel of Experts on Pesticide Residues in Food and the Environment*, considers available data on recognised/registered use patterns, fate of residues, animal and plant metabolism data, analytical methodology and residue data developed through supervised trials. Based on these data, maximum residue levels are proposed for individual pesticides in individual food and feed items or well defined groups of commodities. The deliberations of the JMPR are summarized in he annual Report of the Meeting. The detailed evaluations of the residue (Evaluations

Part II) are also published annually and circulated widely to member governments, international organizations and other interested parties.

OTHER INTER-GOVERNMENTAL ORGANISATIONS

Council of Europe (CoE)

Palais de l'Europe
B.P. 431 R6
F-67006 STRASBOURG Cedex
France
Tel: (88) 614961
Telex: 870943 STRASBOURG

Aims/Activities

CoE has a Committee of Experts on Pesticides which studies the risk to human health arising from the use of pesticides in agriculture and outside agriculture (e.g. in home gardens and food storage), including the study of ways and means of preventing pesticide residues in food as far as possible and to make recommendations which will ensure the safe use of pesticides.

International Standardisation Organisation (ISO)

1 rue de Varembe
Case Postale 56
CH-1211 GENEVA 20
Switzerland
Tel: (22) 341240
Telex: 23887 ISO

Aims/Activities

ISO has a technical committee (ISO/TC81) which sets standards for common names for pesticides and other agrochemicals.

NON-GOVERNMENTAL ORGANISATIONS

Collaborative International Pesticide Analytical Council Limited (CIPAC)

Dr A. Martijn (Secretary)
't Gotink 7
7261 VE RUURLO
The Netherlands
Tel: (573) 452851

Aims/Activities

1. To promote international agreement on:
   - methods for the chemical analysis and for the physico-chemical evaluation of technical pesticide materials and formulations:
   - methods for correlating biological efficacy with physical and chemical properties of pesticide products.
2. To foster inter-laboratory collaborative analysis among interested laboratories.
3. To sponsor symposia to encourage the development of, and to promote discussion on, new or better methods of analysis and physico-chemical evaluation of pesticide products.
4. To publish the methods standardised by the Council, the reports on the work carried out on such agreed methods, the proceedings of symposia and any other appropriate material.

European and Mediterranean Plant Protection Organisation (EPPO)/Working Party on Pesticides for Plant Protection

1 rue Le Notre
F-75016 PARIS
France
Tel: (1) 45207794
Telex: 614148 OEPP
Fax: (1) 42248943

Aims/Activities

1. To achieve active collaboration between Member Countries in order to promote the development and use of a comparable and acceptable approach in the biological evaluation of agricultural pesticides for registration or approval.
2. To exchange information between Member Countries on existing regulations or recommendations on use of pesticides and residue tolerances.
3. To report on special problems encountered in plant protection, in the environment and in trade as a result of use, or restrictions of use, of pesticides.
4. To encourage the use of acceptable alternative materials and methods for control in those instances where this is necessary and possible.
5. To give general recommendations to member Countries on common approaches with respect to minimum requirements in pesticide regulations, particularly where plant protection is involved, taking into account progress made in this field by other international organisations.
6. To collaborate with other international organisations which are currently engaged in the field of pesticides for plant protection.
7. To report annually to the Council of EPPO on the progress made.

International Group of National Associations of Agrochemical Manufacturers (GIFAP)

Avenue Louise 143
B-1050 Brussels
Belgium
Tel: +32 2 542 04 10
Fax: +32 2 542 04 19
E-mail: gifapdg@pophost.eunet.be

Aims/Activities

(a) *To act:*

- on behalf of member associations, to take a position in all technical and scientific domains which interests, directly or indirectly, industry with regard to international and interstate bodies, including the Food and Agriculture Organisation (FAO), World Health Organisation (WHO), the United Nations Environment Programme (UNEP), the United Nations Industrial Development Organisation (UNIDO), the International Labour Organisation (ILO), the World Industrial Property Organisation (WIPO), the World Bank, the Organisation for Economic Cooperation and Development (OECD), the United Nations Educational Scientific Cultural Organisation (UNESCO) and others.
- on behalf of member associations in interpreting the regulatory, scientific and technical recommendations and guidelines issued by international and interstate bodies.

(b) *To promote:*

- the harmonisation of national and international regulations concerning the manufacture, formulation, packing, distribution, labelling, storage, application of crop protection products and their relation with the environment;
- the respect of industrial property rights in the interest of the continuity of scientific and technical progress;
- the extension of the rational use of crop protection products both in the agricultural and vector biology control domains, including the principles of integrated pest management;
- the safe and effective handling and use of crop protection products;
- a better understanding of the progress made with crop protection and public health products, and of the benefits arising from their use.

(c) *To inform:*

- international organisations and the public of industry activities, making known the progress achieved in the domain of crop protection products and their use with respect to agriculture and public health.
- industry, represented by its association members, by disseminating scientific studies, documentation and publications of common interest.

(d) *To support:*
- the policy of member Associations and the crop protection industry by use of modern science and technology to continue to provide crop protection products and technical support to enable an efficient and sustainable world agriculture to produce high quality products meeting the needs of the growing world population, with responsible regard for the environment.
- the activities of member associations of the crop protection industry in communicating the benefits and relative risks of the products of the industry to their members, Governments and the general public.

European Crop Protection Association (ECPA)

Avenue Beaulieu 25
B-1160 BRUSSELS
Belgium
Tel: (2) 66315 60
Fax: (2) 66315 60

Aims/Activities

The European Crop Protection Association was created in June 1992 when it split from the International Group of National Associations of Manufacturers of Agrochemical Products (GIFAP). Its mission is to support the objectives of the crop protection industry, which provides herbicides, fungicides and insecticides for use in agriculture, forestry, horticulture and home gardening. ECPA acts on a pan-European and national basis in the public, legislative, regulatory and social arenas.

Publications

The following ECPA publications are available from the Secretariat. These brochures summarize studies which have been commissioned by the ECPA and carried out by independent consultants, organizations or experts:

- *What the Public Thinks*
  A summary of attitudes to crop protection products and related issues from a survey, conducted by AFC Unternehmensberatung GmbH, in Germany, France, Netherlands, Denmark, and Italy.
- *Organic Farming*
  A summary of findings from a study of seven European countries conducted by the Landell Mills research group.
- *Health Effects in Man from Long Term Exposure to Pesticides*
  A summary of a 1975–1991 literature review, conducted by the International Centre for Pesticide Safety.

- *Cereals and Plant Protection*
  A summary of the cereal chapters, by Dr E.-C. Oerke of the University of Hannover, Germany, from the book '*Protecting the World's Harvest—food needs, crop losses and plant protection*'.
- *Population Growth, Agricultural Production and Food Supplies*
  A summary of the introductory chapters, by A. Weber of the University of Kiel, Germany, from the book '*Protecting the World's Harvest—food needs, crop losses and plant protection*'.
- *What Members of the European Parliament Think*
  A summary of the attitudes of members of the European Parliament towards the Crop Protection Industry and related subjects from survey conducted by Market & Opinion Research International (MOTI).
- *The Public View*
  A summary of attitudes to the crop protection industry, it products and related issues from a survey, conducted by AFC Unternehmensberaturn GmbH in eleven European countries (1993 results).
- *Non-Cereal Crops and Plant Protection*
  A summary of the non-cereal crop chapters, covering cotton, potatoes, soybean and coffee by Dr EC Oerke of the University of Hannover, Germany, from the book '*Protecting the World's Harvest—food needs, crop losses and plant protection*'.
- *Guidelines for the Rinsing of Agrochemical Containers*
  Recommendations based on an international review of rinsing methods.
- *Crop protection in the class room*
  A summary of the teaching materials on crop protection that are currently available to schools and colleges
- *Safe, affordable and sustainable food*
  Selected papers from the 13th International Plant Protection Conference held in The Hague, The Netherlands, July 1995.

In addition to the above publications, the following documentation is available from the Secretariat:
Annual Report, List of Members, List of Publications, ECPA Leaflet.

Pesticides Action Newtork (PAN)

Regional office for North America	Pan North America Regional Centre 116 New Montgomery Street, #810 SANFRANCISCO CA 94105 USA Tel: (415) 5419140 Fax: (415) 5419253 Email: panna@panna.org

Regional office for Latin America	RAPALMIRA Apartado Aéreo 1440 Palmira (Valle) Colombia Tel: (57–227) 35 455 Fax: (57–227) 57 115 Email: rapal@colnodo.apc.org
Regional office for Africa (Anglophone)	Environment Liaison Centre International P.O. Box 72461 NAIROBI Kenya Tel: 254 (2) 562015 Fax: 254 (2) 265175 Email: nkinyua@elci.gn.apc.org
Regional office for Africa (Francophone)	Environnement et Developpement du Tiers Monde B.P. 3370 DAKAR Senegal Tel: 221 225565 Fax: 221 222695 Email: Abou-Thiam@endadak.gn.apc.org
Regional office for Asia/ Pacific	PAN Asia and the Pacific P.O. Box 1170 10850 Penang Malaysia Tel: (60–4) 657 0271 Fax: (60–4) 657 7445 Email: panap@panap.po.my
Regional office for Europe	PAN Europe c/o The Pesticides Trust Eurolink Centre, 49 Effra Road London, SW2 1BZ UK Tel: (44–171) 274 8895 Fax: (44–171) 274 9084 Email: pesttrust@gn.apc.org

Aims/Activities

PAN is a network which has the objective of ending the abuse of pesticides and working towards systems of agriuclture which are non- or less-pesticide-dependent. As a network it has no formal centralised structure, but is co-ordinated regionally in Asia, Africa, North America, Latin America and Europe.

DIRECTORY OF NATIONAL PESTICIDE CONTROL ORGANISATIONS

Compiled by George Ekström, KemI, Solna, Sweden

GUIDE TO THE NATIONAL DIRECTORY

This directory gives contact details of national authorities in 180 countries around the world which are involved in pesticides and their control.

Minimum criterion for inclusion of a country in the directory has been the compiler's awareness of at least one competent authority or other organization that may be used as a contact point. Names of countries have been written in accordance with the 1995 Britannica Book of the Year (Encyclopedia Britannica, Inc).

Drafts of country presentations have been distributed to one or, whenever available, two contact points in each country. The selection of contact points varies from country to country depending on infrastructure.

The contents of the country presentations have been standardized in a total of thirteen fields. The first seven fields cover national domestic activities performed by competent authorities, research institutions, laboratories etc. responsible for various aspects of the national government's pesticide control strategy.

The 8th field covers the Designated National Authority (DNA) acting as liaison with the Food and Agriculture Organization (FAO), and the International Register of Potentially Toxic Chemicals (UNEP/IRPTC) concerning notifications and information exchange regarding banned and severely restricted pesticides based on the London Guidelines and the Prior Informed Consent (PIC) procedures. (Countries may have designated the same or a different authority for consumer and industrial chemicals.)

The 9th field covers the national contact point of the Codex Alimentarius Commission of the Joint FAO/WHO Food Standards Programme. The Codex Alimentarius' Committee on Pesticide Residues (CCPR) recommends for international use maximum residue limits (MRLs) for pesticide residues in food and animal feed.

The 10th field covers national correspondents to the International Register of Potentially Toxic Chemicals (UNEP/IRPTC) in Geneva.

The 11th field covers national focal points for Infoterra, the international environmental information system set up by the United Nations Environment Programme (UNEP) in Nairobi.

The 12th field covers designated focal points for the Inter-Governmental Forum on Chemical Safety (IFCS), established at the International Conference on Chemical Safety held in Stockholm (Sweden) in April 1994.

The 13th field covers national (non-governmental) associations of national pesticide importers, manufacturers, formulators and distributors.

In many cases a 14th field may be found at the bottom of the page stating national Poisons Information/Control centres, if available.

References to responsible authorities and other organisations have been given with one or more numbers. A dash (-) denotes the non-existence of a competent authority or other organisation, whereas an empty space in the same column denotes unavailable information.

Albania
Republika e Shqipérisé

Responsibility	
Main authority responsible for approval/registration of pesticides	3
Main authority for pesticide use directives, permits and training	12
Main authority responsible for establishing national maximum pesticide residue limits	2, 4, 5, 6, 12
Main authority responsible for control of pesticide residues in food	2,5
Main laboratory for analysis of pesticide residues in food	4,6
Main laboratory for quality control of pesticide products	7
Main organisation responsible for applied research regarding pesticide use. Information on related regional research	7
Designated national authority for notifications and information regarding banned and severely restricted pesticides including the PIC procedure	1, 12
Codex Alimentarius contact point	8
National IRPTC correspondent	10
Infoterra national focal point	11
National IFCS focal point	9
National association of pesticide manufacturers	-

Responsible organisations

1. CEP - Committee of Environmental Protection
   Ministry of Health and Environmental Protection
   TIRANA
   Tel: (355 42) 27907
   Fax: (355 42) 27907

2. Primary Health Care Department
   Ministry of Health and Environmental Protection
   TIRANA
   Tel: (355 42) 24671

3. State Commission of Pesticides
   Ministry of Agriculture and Food
   TIRANA
   Tel: (355 42) 29309

4. Institute of Public Health
   TIRANA
   Tel: (355 42) 33553
   Fax: (355 42) 33553

5. Directory for Food Insurance
   Ministry of Agriculture and Food
   TIRANA
   Tel: (355 42) 22682

6. Food Technology Institute
   TIRANA
   Tel: (355 42) 26770
   Fax: (355 42) 26700

7. Plant Protection Institute
   Shkozet
   DURRES
   Tel: (355 52) 22182
   Fax: (355 52) 22182

8. Primary Health Care Department
   Attn: Dr. Agim Shehi
   Ministry of Health and Environmental Protection
   TIRANA

9. Department of Public Health
   Attn: Mrs. Raimonda Totoni
   Ministry of Health and Environmental Protection
   TIRANA
   Tel: (355 42) 32731
   Fax: (355 42) 33553

10. Dr. G Gjyli
    Director,
    Research Institute of Chemical Technology and Design
    Block Vasil Shanto
    TIRANA
    Tel: (355 42) 23466
    Fax: (355 42) 23466

11. Infoterra NFP
    Environmental Department
    Centre for Scientific and Technical Information and Documentation (CSTID)
    Committee of Science and Technology
    Rr. Leke Dukagjini, 5
    TIRANA
    Tel: (355 42) 22491
    Fax: (355 42) 27975

12. Directorate of Plant Protection Service and Plant Protection
    Bulevardi "Deshmoret e Kombit"
    Ministry of Agriculture and Food
    TIRANA
    Tel: (355 42) 29309
    Fax: (355 42) 27924

Algeria
al-Jumhuriyah al-Jazairiyah ad-Dimuqratiyah ash-Shabiyah

Responsibility

Main authority responsible for approval/registration of pesticides	1
Main authority responsible for pesticide use directives, permits and training	
Main authority responsible for establishing national maximum pesticide residue limits	1, 2, 3
Main authority responsible for control of pesticide residues in food	1, 3
Main laboratory for analysis of pesticide residues in food	-
Main laboratory for quality control of pesticide products	
Main organisation responsible for applied research regarding pesticide use. Information on related regional research	1
Designated national authority for notifications and information regarding banned and severely restricted pesticides including the PIC procedure	1, 4
Codex Alimentarius contact point	5
National IRPTC correspondent	4, 6
Infoterra national focal point	7
National IFCS focal point	-
National association of pesticide manufacturers	-

Notes

Poisons information/control centres in Algiers and Constantine

Responsible organisations

1. Ministère de l'Agruculture
   Institut National de la Protection des Vegetaux (INPV)
   B.P. 80
   El Harrach
   ALGER
   Tel: (213 2) 523 016
   Fax: (213 2) 766 271, 525 863

2. Ministère de la Santé Publique
   Centre de la Toxicologie
   Centre de l'Hôpital Universitaire
   Mustapha
   Place de ler Mai
   ALGER
   Tel: 663727, 663729

3. Ministère du Commerce
   Laboratoire Centrale du Contrôle de la Qualité
   Chemin rural no 8, Houch Hadda
   Hassan Badi
   El Harrach
   ALGER
   Tel: 766140, 764243

4. Ministère de l'intérieur, des collectivités locales, de l'environnement et de la réforme administrative
   Direction generale de l'environnement
   Direction de la prévention de pollutions et nuisances
   Palais du Gouvernement
   17 Bd. Docteur Saadane
   ALGER
   Tel: 590 205
   Fax: 605 072

5. Ministère de l'Economie
   Direction Générale de la Concurrence et des Prix
   Direction de la Qualité et de la Consommation
   Palais du Gouvernement
   ALGER

6. Ministère de l'Enseignement Supérieur
   Direction de la Post-Graduation et de la Recherche Scientifique
   Route de Deley Brahim
   Ben-Aknoun
   ALGER

7. M. M. Abdelhalim
   Point Focal National d'Infoterra
   Agence Nationale pour la Protection de l'Environnement (ANPE)
   B.P. 62 Bordj el Bahri
   W. Boumerdes
   ALGER
   Tel: (213 2) 861449, 863773
   Fax: (213 2) 863775

Angola
República de Angola

Responsibility	
Main authority responsible for approval/registration of pesticides	1
Main authority responsible for use directives, permits and training	
Main authority responsible for establishing national maximum pesticide residue limits	-
Main authority responsible for control of pesticide residues in food	2
Main laboratory for analysis of pesticide residues in food	2
Main laboratory for quality control of pesticide products	
Main organisation responsible for applied research regarding pesticide use. Information on related regional research	2
Designated national authority for notifications and information regarding banned and severely restricted pesticides including the PIC procedure	3
Codex Alimentarius contact point	4
National IRPTC correspondent	-
Infoterra national focal point	5
National IFCS focal point	6
National association of pesticide manufacturers	-

Responsible organisations

1. Ministerio da Agricultura
Plant Protection Service
(DNOPA)
Cx. Postal No. 527
LUANDA

2. Direccáo-Geral de Proteccáo
da Producáo Agrícola
(DGPPA)
Quinta da Marqués; P-2780
OEIRAS
Portugal
Tel: 2430772

3. Le Coordinateur
Programme National de la
Protection des Plantes
Ministère de l'Agriculture
Cabinet Technique
Avenida Cdt. Gika
LUANDA
Tel: 32357, 321 568

4. Mr. Estevão Miguel de
Carvalho
Director, Gabinete Tecnico
Ministerio da Agricultura e
Desenvolvimento Rural
C.P. 527
LUANDA

5. Mr. Antonio M.F. Bartolomeu
Infoterra NFP
State Secretariat of
Environment
LUANDA
Tel: (244 2) 335844, 335893
Fax: (244 2) 332611

6. Mr David N'Gove Lussoke
Ministerio da Administracao
Publica Emprego e
Seguranea Social
Departement National de la
Securite et de l'Hygiene au
Travail
C.P.6967
LUANDA
Tel:(224-2) 392 381
Fax: (224-2) 393 384

Antigua and Barbuda

Responsibility	
Main authority responsible for approval/registration of pesticides	4
Main authority responsible for use directives, permits and training	-
Main authority responsible for establishing national maximum pesticide residue limits	1, 4
Main authority responsible for control of pesticide residues in food	1, 5
Main laboratory for analysis of pesticide residues in food	3
Main laboratory for quality control of pesticide products	-
Main organisation responsible for applied research regarding pesticide use. Information on related regional research	1, 2, 6, 7
Designated national authority for notifications and information regarding banned and severely restricted pesticides including the PIC procedure	4
Codex Alimentarius contact point	1
National IRPTC correspondent	-
Infoterra national focal point	8
National IFCS focal point	5
National association of pesticide manufacturers	-

Responsible organisations

1. The Director,
   Antigua and Barbuda
   Bureau of Standards (ABBS)
   P.O Box 1550
   ST. JOHN'S
   Tel: (809) 462 1542
   Fax: (809) 462 1625

2. The Director
   Ministry of Agriculture,
   Fisheries, Lands and
   Housing
   Temple Street
   ST. JOHN'S
   Tel: (809) 462 4962
   Fax: (809) 462 6104

3. Chief Government Chemist
   Chemistry and Food
   Technology
   Dunbars
   Friars Hill Road
   ST. JOHN'S
   Tel: (809) 462 6281
   Fax: (809) 462 4502

4. Pesticide Control Board
   Ministry of Agriculture
   Nevis and Temple Streets
   ST. JOHN'S
   Tel: (809) 462 4962
   Fax: (809) 462 6104

5. Chief Health Inspector
   Ministry of Health and Home
   Affairs
   St John's Health Centre
   All Saints Road
   ST. JOHN'S
   Tel: (809) 462 1891, 462 2936
   Fax: (809) 462 1595

6. CEHI
   P.O. Box 1111
   Morne, Castries
   ST. LUCIA
   Tel: (809) 452 2501
   Fax: (809) 453 2721

7. CARDI
   University Campus
   ST. AUGUSTINE
   Trinidad W.I.
   Tel: (809) 645 1205
   Fax: (809) 645 1208

8. Mrs. Candia Williams
   Infoterra NFP
   Ministry of Tourism, Culture
   and Environment
   Queen Elizabeth Highway
   ST. JOHN'S
   Tel: (1 809) 462 4625
   Fax: (1 809) 462 2836

Argentina
República Argentina

Responsibility	
Main authority responsible for approval/registration of pesticides	1, 2
Main authority responsible for use directives, permits and training	
Main authority responsible for establishing national maximum pesticide residue limits	3
Main authority responsible for control of pesticide residues in food	4, 5, 6
Main laboratory for analysis of pesticide residues in food	7, 8
Main laboratory for quality control of pesticide products	
Main organisation responsible for applied research regarding pesticide use. Information on related regional research	9, 10
Designated national authority for notifications and information regarding banned and severely restricted pesticides including the PIC procedure	11
Codex Alimentarius contact point	13
National IRPTC correspondent	14
Infoterra national focal point	15
National IFCS focal point	12
National association of pesticide manufacturers	16

Note

Poisons information/control centres in Buenos Aires and Santa Fe

Responsible organisations

1. Ministerio de Salud y Acción Social
Defensa 120
BUENOS AIRES

2. Secretaría de Comercio de la Nación
Avenida Julio A Roca 651
BUENOS AIRES
Tel: 307241, 303970

3. Secretaría de Agricultura y Ganadería de la Nación
Paseo Colón 922/982 Cod. 1305
BUENOS AIRES
Tel: 3627111, 3627213, 3627281

4. Instituto Nacional de Tecnología Industrial (INTI)
Leandro N. Alem 1067
BUENOS AIRES
Tel: 3133013, 3130931

5. Dirección de Química
Madero 279
BUENOS AIRES
Tel: 333263, 336061

6. Instituto Nacional de Farmacología y Bromatología
Caseros 2170
BUENOS AIRES
Tel: 279191

7. Servicio Nacional de Laboratorios de Productos Ganaderos
Paseo Colón 922–2°p., Of. 230
BUENOS AIRES
Tel: 3622788

8. Instituto Nacional de Tecnología Agropecuaria (INTA)
Avda. Rivadavia 1439 1°p
BUENOS AIRES
Tel: 37095

9. Comisión Investigaciones Científicas de la Provincia de Buenos Aires
Calle 526 entre 10 y 11
LA PLATA

10. Universidad de Buenos Aires
Viamonte 444
BUENOS AIRES

11. Director General del Instituto Argentino de Sanidad y Calidad Vegetal
Ing. Huergo No. 1001
C.P. 1060 BUENOS AIRES
Fax: 541 1615

12. Dr Jorge Herkovists
Director,
Instituto de Ciencias Ambientales y Salud
Programa de Seguridad Quimica
Nicasio Orono 710
1405 CAPITAL FEDERAL
Tel: (54 1) 431 2445
Fax: (54 1) 431 4206

13. Dr A M Sanchez
Coordinador del Codex Alimentarius
Secretaría de Estado de Comercio y Negociaciones Económicas Internacionales
Avenida Julio A Rocca 651, 5° piso, Sector 13
1000 BUENOS AIRES

14. Director,
Direccion de Calidad Ambiental
Secretaría de Vivienda y Calidad Ambiental
Ministerio de Salud y Action Social
Avda. 9 de Julio 1925, piso 17
C.P. 1332
BUENOS AIRES

15. Dra. Marta Zas
CNC Infoterra
Secretaria de Recursos Naturales y Ambiente Humano
San Martin 459, Piso 1
Capital Federal 1004
BUENOS AIRES
Tel: (54-1) 322 4778

16. Cámara de Sanidad Agropecuaria y Fertilizantes (CASAFE)
Rivadavia 1367,7 piso B
1033 BUENOS AIRES
Tel: (54 1) 381 2742,
311 8504
Fax: (54 1) 383 1562

Armenia
Hayastani Hanrapetut'yun

Responsibility	
Main authority responsible for approval/registration of pesticides	2, 5
Main authority responsible for pesticide use directives, permits and training	2, 5
Main authority responsible for establishing national maximum pesticide residue limits	2, 5
Main authority responsible for control of pesticide residues in food	4
Main laboratory for analysis of pesticide residues in food	3, 4, 7
Main laboratory for quality control of pesticide products	7
Main organisation responsible for applied research regarding pesticide use. Information on related regional research	4, 9
Designated national authority for notifications and information regarding banned and severely restricted pesticides including the PIC procedure	4, 6
Codex Alimentarius contact point	8
National IRPTC correspondent	10
Infoterra national focal point	1
National IFCS focal point	-
National association of pesticide manufacturers	-

Responsible organisations

1. Ministry of Econonomy
Armenian Institute of Scientific and Technical Information (ARMNIINTI)
Infoterra NFP
Komitasa Street 49/3
YEREVAN 375051
Tel: (7 8852)236 774, 238 027

2. Ministry of Health
Hygiene and Epidemic Control Administration
Tumanyar Street 8
YEREVAN 375001
Tel: (7 8852) 520 660
Fax: (7 8852) 520 660

3. Ministry of Health
Republican Centre of Hygiene and Epidemic Control
David Malyan Street 37
YEREVAN 375096
Tel: (7 8852) 613 393
Fax: (7 8852) 621 336

4. Ministry of Health
Scientific Research Institute of Environmental Hygiene and Prophylactic Toxicology
Halabyar Street 46
YEREVAN 375108
Tel: (7 8852) 358 235

5. Ministry of Agriculture and Food
"ArmFertility" (HayBeriutyun) Concern
Amaranotzain Street 125
YEREVAN 375047
Tel: (7 8852)520 293, 653 500

6. Ministry of Agriculture and Food
"ArmFertility"
(HayBerriutyun) Concern
Plant Protection Service
Amaranotzain Street 125
YEREVAN 375047
Tel: (7 8852)651 911

7. Ministry of Agriculture and Food
"ArmFertility"
(HayBerriutyun) Concern
Plant Protection and Control Service
Toxicological Laboratory
Isakova Street 22
YEREVAN 375082
Tel: (7 8852) 424 821

8. Ministry of Agriculture and Food
"Paren" (Food) Scientific, Productive and Project State Company
Gorvetka Street 4
YEREVAN 375023
Tel: (7 8852) 524 687

9. Ministry of Agriculture and Food
Scientific Research Institute of Plant Protection
Village Merdzavan
ETCHMIADZIN 378312

10. Ministry of Nature Protection
Moscovian Street 35
YEREVAN 375002
Tel: (7 8852) 538 613
Fax: (7 8852) 534 902

Australia

Commonwealth of Australia

Responsibility	
Main authority responsible for approval/registration of pesticides	1
Main authority responsible for pesticide use directives, permits and training	
Main authority responsible for establishing national maximum pesticide residue limits	2
Main authority responsible for control of pesticide residues in food	3
Main laboratory for analysis of pesticide residues in food	4
Main laboratory for quality control of pesticide products	
Main organisation responsible for applied research regarding pesticide use. Information on related regional research	5
Designated national authority for notifications and information regarding banned and severely restricted pesticides including the PIC procedure	6
Codex Alimentarius contact point	7
National IRPTC correspondent	8
Infoterra national focal point	9
National IFCS focal point	11
National association of pesticide manufacturers	10

Note

Poisons information/control centres in Adelaide, Brisbane, Canberra, Darwin, Hobart, Melbourne, Perth and Sydney

Responsible organisations

1. National Registration Authority for Agricultural and Veterinary Chemicals
GPO Box 858
CANBERRA ACT 2600

2. National Health and Medical Research Council
GPO Box 9848
CANBERRA ACT 2601
Tel: 62 891 555
Fax: 62 816 946

3. Department of Primary Industries and Energy
GPO Box 858
CANBERRA ACT 2600
Tel: 62 716 370
Telex: 62188

4. Australian Government Analytical Laboratories
P.O. Box 65
BELCONNEN ACT 2616

5. Division of Plant Industry
Commonwealth Scientific Organization (CSIRO)
P.O. Box 1600
BLACK MOUNTAIN ACT 2365

6. Director,
Agricultural and Veterinary Chemicals Policy Section
Department of Primary Industries and Energy
GPO Box 858
CANBERRA ACT 2601
Tel: (61 6) 271 6371
Fax: (61 6) 272 5899

7. Mr. D. Gascoine
Director, Food Inspection Division
Australian Quarantine and Inspection Service
GPO Box 858
CANBERRA ACT 2601
Tel: 6 27 25 584
Fax: 6 27 25 226

8. The Director,
Chemical Policy Section
Commonwealth Environment Protection Agency (CEPA)
Attn: Dr. Geoff Thompson
Department of the Arts, Sport, the Environment, Tourism and Territories
P.O. Box E305
Queen Victoria Terrace
BARTON ACT 2600

9. Ms. Kay L. Winter
Infoterra National Focal Point
Department of the Environment, Sport and Territories
GPO Box 787
CANBERRA ACT 2601
Tel: (61 62) 741 209
Fax: (61 62) 741 215

10. The Agricultural and Veterinary Chemicals Association of
Australia (AVCA)
Private Bag 938
NORTH SYDNEY, NSW 2059
Tel: (61 2) 963 7690
Fax: (61 2) 929 0213

11. Ms Kaye Dal Bon,
Assistant Secretary
Environment Standards
Branch
Environment Protection
Agency
CANBERRA ACT 2601
Fax: (61 6) 274 1172

Austria
Republik Österreich

Responsibility

Main authority responsible for approval/registration of pesticides	1, 2, 3, 8
Main authority responsible for pesticide use directives, permits and training	
Main authority responsible for establishing national maximum pesticide residue limits	1, 3
Main authority responsible for control of pesticide residues in food	4
Main laboratory for analysis of pesticide residues in food	2, 4, 5
Main laboratory for quality control of pesticide products	
Main organisation responsible for applied research regarding pesticide use. Information on related regional research	2, 12
Designated national authority for notifications and information regarding banned and severely restricted pesticides including the PIC procedure	8
Codex Alimentarius contact point	9
National IRPTC correspondent	-
Infoterra national focal point	10
National IFCS focal point	11
National association of pesticide manufacturers	6

Note

Poisons information/control centre: Vergiftungsinformationszentrale, Allgemeines Krankenhaus, Wahringer Gurtel 18-20, 1090 Wien, Tel: (43 222) 40400-2222, Fax: (43 222) 40400-4225 [Head: Dr. K. Hruby]

Responsible organisations

1. Bundesministerium für Land- und Forstwirtschaft
Abteilung VI/C9
Stubenring 1
A-1010 WIEN
Tel: (43 1) 71100-6802
Fax: (43 1) 71100-2901

2. Bundesanstalt für Pflanzenschutz
Spargelfeldstrasse 191
A-1220 WIEN
Tel: (43 1) 288-16-0
Fax: (43 1) 288-16/5216

3. Bundesministerium für Gesundheit und Konsumentenschutz
Radetzkystrasse 2
A-1030 WIEN
Tel: (43 1) 711 720
Fax: (43 1) 713 7952

4. Bundesanstalt für Lebensmitteluntersuchung und -forschung
Kinderspitalgasse 15
A-1090 WIEN
Tel: (43 1) 404 910
Fax: (43 1) 40491 540

5. Lebensmittelversuchsanstalt
Blaasstrasse 29
A-1190 WIEN
Tel: (43 1) 368-85-550
Fax: (43 1) 3688/555

6. Fachverband der Chemischen Industrie Österreichs (FCIÖ)
Wiedner Hauptstrasse 63
P.O. Box 325
A-1045 WIEN
Tel: (43 1) 5010 53371
Fax: (43 1) 5020 6280

7. Universität für Bodenkultur
Institut für Pflanzenschutz
Peter Jordanstrasse 82
A-1190 WIEN
Tel: (43 1) 47654 3350
Fax: (43 1) 310 5177

8. Bundesministerium für Umwelt, Jugend und Familie
Dep. II/3
Stubenbastei 5
A-1010 WIEN
Tel: (43 1) 51522-2701
Fax: (43 1) 51522-7744

9. Bundesministerium für Land- und Forstwirtschaft
(Dep. III/3)
Stubenring 12
A-1010 WIEN
Tel: (43 1) 71100-2812
Fax: (43 1) 71100-2892

10. Dr. Johannes Mayer
Umweltbundesamt
Austrian Infoterra NFP
Spittelauer Laende 5
A-1090 WIEN
Tel: (43 1) 31304-3240
Fax: (43 122) 31304-5400

Responsible organisations

11. Federal Ministry for the Environment
    Department II/3
    Stubenbaster 5
    A-101 WIEN
    Tel: (43 1) 531522 2702
    Fax: (43 1) 51522 7744

12. Bundesamt und Forschungszentrum fur Landwirtschaft
    Spergelfeldstrasse 191
    A-1220 WIEN
    Tel: (43 1) 28816-0
    Fax: (43 1) 28816-5194

Azerbaijan

Azärbayjan Respublikasi

Responsibility

Main authority responsible for approval/registration of pesticides	
Main authority responsible for pesticide use directives, permits and training	
Main authority responsible for establishing national maximum pesticide residue limits	
Main authority responsible for control of pesticide residues in food	
Main laboratory for analysis of pesticide residues in food	
Main laboratory for quality control of pesticide products	
Main organisation responsible for applied research regarding pesticide use. Information on related regional research	
Designated national authority for notifications and information regarding banned and severely restricted pesticides including the PIC procedure	-
Codex Alimentarius contact point	-
National IRPTC correspondent	1
Infoterra national focal point	2
National IFCS focal point	-
National association of pesticide manufacturers	-

Responsible organisations

1. Deputy Main State Sanitary Physician
   Dr. V K Kasumov
   State Committee on Ecology and Control of Nature Utilization
   Istiglaliyyat Str. 31
   370001 BAKU

2. Prof. D. Mekhtiev
   Director General,
   Information Union AZNTI
   370005 GSP 601
   Natavan Line 2/6 Rio "AZNTI"
   BAKU
   Tel: (7 8922) 933341

Bahamas

Commonwealth of the Bahamas

Responsibility	
Main authority responsible for approval/registration of pesticides	1
Main authority responsible for pesticide use directives, permits and training	1
Main authority responsible for establishing national maximum pesticide residue limits	1
Main authority responsible for control of pesticide residues in food	1
Main laboratory for analysis of pesticide residues in food	1
Main laboratory for quality control of pesticide products	-
Main organisation responsible for applied research regarding pesticide use. Information on related regional research	1
Designated national authority for notifications and information regarding banned and severely restricted pesticides including the PIC procedure	1
Codex Alimentarius contact point	1
National IRPTC correspondent	3
Infoterra national focal point	4
National IFCS focal point	2
National association of pesticide manufacturers	-

Responsible organisations

1. Director of Agriculture
   Ministry of Agriculture, Trade and Industry
   P.O. Box N-3028
   NASSAU

2. Director of Environmental Health Services
   Ministry of Health and Environment
   P.O. Box N-3729
   NASSAU
   Tel: (809) 322 4908
   Fax: (890) 322 3607

3. Chief Medical Officer
   Ministry of Health and Environment
   P.O. Box N-3730
   NASSAU

4. The Chairman,
   Bahamas Environment, Science and Technology Commission
   Office of the Prime Minister
   P.O. Box CB-10980
   NASSAU
   Tel: (1 809) 327 4691, 327 5826
   Fax: (1 809) 327 4626, 327 5806

Bahrain
Dawlat al-Bahrayn

Responsibility	
Main authority responsible for approval/registration of pesticides	1, 7
Main authority responsible for pesticide use directives, permits and training	1, 4
Main authority responsible for establishing national maximum pesticide residue limits	3
Main authority responsible for control of pesticide residues in food	3
Main laboratory for analysis of pesticide residues in food	2
Main laboratory for quality control of pesticide products	-
Main organisation responsible for applied research regarding pesticide use. Information on related regional research	-
Designated national authority for notifications and information regarding banned and severely restricted pesticides including the PIC procedure	8
Codex Alimentarius contact point	3
National IRPTC correspondent	6
Infoterra national focal point	5
National IFCS focal point	4
National association of pesticide manufacturers	-

Responsible organisations

1. Mr. Hassan Abdul Karim
Director of Agriculture
Ministry of Works and Agriculture
P.O. Box 251
MANAMA
Fax: 695527

2. Public Health Laboratory
P.O. Box 42
MANAMA
Tel: 255555
Telex: 8511 health bn

3. Dr. Sameer Khalfan
Ag. Director of Public Health
Ministry of Health
P.O. Box 12
MANAMA
Tel: 250313
Fax: 279253

4. Mrs. Affaf Al-Shoala
Occupational Health Specialist
Ministry of Health
P.O. Box 42
MANAMA
Tel: 279224
Fax: 258221

5. Mr. Khalid Fakhro
Vice Chairman,
Environmental Protection Committee
P.O. Box 26909
MANAMA
Tel: 293693
Fax: 293694

6. H.E. Sh. Khalid Bin Abdulla Al-Khalifa
Minister of Housing and Environment
P.O. Box 802
MANAMA
Tel: 528200
Fax: 534115

7. Ministry of Health
P.O. Box 12
MANAMA
Tel: 255555
Fax: 252569

8. H.E.Mr. Jawed Salim Al-Arayed,
Minister of Health
Ministry of Health
P.O. Box 12
MANAMA
Tel: (973) 252 605
Fax: (973) 252 569

Bangladesh

Gana Prajatantri Bangladesh

Responsibility	
Main authority responsible for approval/registration of pesticides	1, 2
Main authority responsible for pesticide use directives, permits and training	
Main authority responsible for establishing national maximum pesticide residue limits	3
Main authority responsible for control of pesticide residues in food	1, 3, 4
Main laboratory for analysis of pesticide residues in food	1
Main laboratory for quality control of pesticide products	
Main organisation responsible for applied research regarding pesticide use. Information on related regional research	1, 5, 6, 7
Designated national authority for notifications and information regarding banned and severely restricted pesticides including the PIC procedure	8
Codex Alimentarius contact point	9
National IRPTC correspondent	13
Infoterra national focal point	10
National IFCS focal point	12
National association of pesticide manufacturers	11

Responsible organisations

1. Department of Environment Pollution Control (DEPC)
6/11/F Lalmatia Housing Estate
Satmasjid Road
DHAKA 7

2. Pesticide Technical Advisory Committee
The Secretary, Ministry of Agriculture
Bangladesh Secretariat
DHAKA

3. Central Testing Laboratory (CTL)
Tejgaon Industrial Area
DHAKA

4. Plant Protection Wing
Khamarbari, Farmgate
DHAKA

5. Bangladesh Agriculture University
MYMENSINGH

6. Bangladesh Agricultural Research Institute
Joydebpur
DHAKA

7. Bangladesh Rice Research Institute
Joydebpur
DHAKA

8. The Director General
Deparment of Environment
House No. 2, Road No. 16
Dhanmondi R/A
DHAKA 1209
Tel: 315 777, 318 682

9. The Director-General
Bangladesh Standards and Testing Institution (BSTI)
116/A, Tejgaon Industrial Area
DHAKA 8

10. Mr. Syed A N M Wahed
Infoterra NFP
Department of Environment
Government of the People's Republic of Bangladesh
House No. 2, Road No. 16 (new)
Dhanmondi
DHAKA 12091
Tel: (880 2) 312 489, 318 682

11. Pesticide Association of Bangladesh (PAB)
c/o Ciba Geigy Bangladesh Ltd.
House No. 50,
Road No. 2 A
Dhanmondi R/A
G.P.O. Box 431
DHAKA 1209
Tel: (880 2) 865 302, 500 1117
Fax: (880 2) 863 489

12. Mr Mohammad Mosharaf Hossain
Deputy Director,
Department of Environment
House No.2, Road No.16
Dhanmondi Residential Area
DHAKA

13. The Director,
Environment Pollution Control (EPC)
6/11/F Lalmatia Housing Estate
Satmasjid Rd.
DHAKA 7

Barbados

Responsibility	
Main authority responsible for approval/registration of pesticides	1
Main laboratory responsible for pesticide use directives, permits and training	2
Main authority responsible for establishing national maximum pesticide residue limits	3
Main authority responsible for control of pesticide residues in food	3
Main laboratory for analysis of pesticide residues in food	6
Main laboratory for quality control of pesticide products	2
Main organisation responsible for applied research regarding pesticide use. Information on related regional research	4
Designated national authority for notifications and information regarding banned and severely restricted pesticides including the PIC procedure	2
Codex Alimentarius contact point	3
National IRPTC correspondent	4
Infoterra national focal point	5
National IFCS focal point	-
National association of pesticide manufacturers	-

Responsible organisations

1. Chief Agricultural Officer
   Ministry of Agriculture and
   Rural Development
   P.O. Box 505
   Graeme Hall
   CHRISTCHURCH
   Tel: (809) 428 4060, 428 4150
   Fax: (809) 420 8444

2. The Deputy Chief
   Agricultural Officer
   Ministry of Agriculture and
   Rural Development
   P.O. Box 505
   Graeme Hall
   CHRISTCHURCH
   Tel: (809) 428 4150
   Fax: (809) 420 8444

3. The Director
   Barbados National Standards
   Institution
   "Flodden",
   Culloden Road
   ST. MICHAEL
   Tel: (809) 426 3870
   Fax: (809) 436 1495

4. The Chief Medical Officer
   Ministry of Health and the
   Environment
   Jemmots Lane
   ST. MICHAEL
   Tel: (809) 426 3470
   Fax: (809) 426 5570

5. The Permanent Secretary
   (Environment Division)
   Ministry of Health and the
   Environment
   4th Floor,
   Sir Frank Walcott Building
   Culloden Road
   ST. MICHAEL
   Tel: (809) 431 7680
   Fax: (809) 437 8859

6. The Director
   Government Analytical
   Services Laboratory
   Culloden Road
   ST. MICHAEL
   Tel: (809) 426 0683
   Fax: (809) 436 7682

Belarus
Respublika Belarus

Responsibility

Main authority responsible for approval/registration of pesticides	1, 2, 4
Main authority responsible for pesticide use directives, permits and training	2
Main authority responsible for establishing national maximum pesticide residue limits	1
Main authority responsible for control of pesticide residues in food	1
Main laboratory for analysis of pesticide residues in food	5
Main laboratory for quality control of pesticide products	6
Main organisation responsible for applied research regarding pesticide use. Information on related regional research	6, 9
Designated national authority for notifications and information regarding banned and severely restricted pesticides including the PIC procedure	3
Codex Alimentarius contact point	-
National IRPTC correspondent	-
Infoterra national focal point	7
National IFCS focal point	8
National association of pesticide manufacturers	-

Note

Poisons information/control centre: Minsk Municipal Hospital, Kizhevatova Str 58, Minsk, Tel: (7) 78 81 21 [Head: Dr H Karpuk]

Responsible organizations

1. Ministry of Health
   Ulitsa Mjasnikova 39
   220097 MINSK
   Fax: 296 297

2. Ministry of Agriculture and Food
   Ulitsa Kirova 15
   220031 MINSK
   Fax: 274 388

3. Republican Centre for Hygiene and Epidemiology
   Ministry of Health
   Ulitsa Kazintsa 50
   220108 MINSK
   Fax: 784 226

4. State Committee for Registration of Plant Protection Products
   Ministry of Agriculture and Food
   Ulitsa Kazintza 62
   220108 MINSK
   Fax: 274 388

5. Laboratory for Control and Toxicology
   Ministry of Agriculture and Food
   Ulitsa Kommunalnaja 9
   P/O Sennitsa
   223056 MINSKIJ R-N
   Fax: 274 388

6. Research Institute for Plant Protection
   Ministry of Agriculture and Food
   P/O Priluki
   223011 MINSKIJ R-N
   Fax: 992 338

7. Mr. A. Aleshka
   Infoterra NFP
   Director,
   Byelorussian Research Centre (Ecology)
   31A, Horuzhaya Street
   220002 MINSK
   Tel: (7 0172) 347 065
   Fax: (7 0172) 348 072

8. Dr. Igor A Michailopulo
   Deputy Director,
   Bio-organic Chemistry Institute of the Academy of Sciences
   5 Xhodinskaya Street
   220733 MINSK
   Tel: (7 0172) 648 761
   Fax: (7 0172) 636 772

9. Territorial Centres for Hygiene and Epidemiology
   c/o Ministry of Health
   Minsk Regional Centre
   Ulitsa P Brovki
   MINSK
   Tel: 314 685

Belgium

Royaume de Belgique, Koninkrijk België

Responsibility	
Main authority responsible for approval/registration of pesticides	1
Main authority responsible for pesticide use directives, permits and training	
Main authority responsible for establishing national maximum pesticide residue limits	1, 8
Main authority responsible for control of pesticide residues in food	2
Main laboratory for analysis of pesticide residues in food	3
Main laboratory for quality control of pesticide products	
Main organisation responsible for applied research regarding pesticide use. Information on related regional research	4
Designated national authority for notifications and information regarding banned and severely restricted pesticides including the PIC procedure	6
Codex Alimentarius contact point	7
National IRPTC correspondent	8
Infoterra national focal point	9
National IFCS focal point	10
National association of pesticide manufacturers	5

Note

Poisons information/control centre: Centre Anti-Poisons, Rue Joseph Stallaert 1, Boite 15, 1060 Bruxelles, Tel: (32 2) 345 1818, Fax: (32 2) 347 5860 [Head: Dr. Ms. M. Mostin]

Responsible organisations

1. Comité d'Agréation des Produits Phytopharmaceutiques
   Inspection des Matières Premières
   Ministère de l'Agriculture
   Manhattan Center
   21 Avenue du Boulevard
   B-1210 BRUXELLES
   Tel: (2) 211 7211
   Fax: (2) 211 7216

2. Ministère de la Santé Publique et de la Famille
   Inspection des Denrées Alimentaires
   Cité Administrative de l'État
   4 Quartier Vésale
   B-1010 BRUXELLES
   Tel: (2) 210 4858
   Telex: 25768 MUGSPR

3. Institut d'Hygiène et d'Épidémiologie
   Analyse des Denrées Alimentaires
   Ministère de la Santé Publique et de la Famille
   14 rue Juliette Wytsman
   B-1050 BRUXELLES
   Tel: (2) 647 9980
   Fax: (2) 642 5001

4. Station de Phytopharmacie de l'État
   Centre de Recherches Agronomiques de l'État
   Ministère de l'Agriculture
   11 rue du Bordia
   B-5800 GEMBLOUX
   Tel: (81) 612971
   Fax: (81) 614941

5. Association Belge de l'Industrie des Produits Phytosanitaires (PHYTOPHAR)
   49 Square Marie-Louise
   B-1040 BRUXELLES
   Tel: (32 2) 238 9711, 339 711
   Fax: (32 2) 238 1301, 231 1301

6. Service Maitrise des risques
   Section Pesticides (bureau 2/309)
   Ministère de la Santé Publique et de l'Environnement
   Cité Administrative de l'État
   B-1010 BRUXELLES
   Tel: (32 2) 210 4881
   Fax: (32 2) 210 4884

7. Comité Belge du Codex Alimentarius
   Service du Commerce International des Matières Premiéres et Produits Agricoles (B14)
   Ministère des Relations Extérieures
   2 rue Quatre-Bras
   B-1000 BRUXELLES
   Tel: 516 8299
   Fax: 516 8827

8. Chef de la Section Toxicologie
   Institut d'Hygiène et d'Épidémiologie
   Ministère de la Santé Publique et de l'Environnement
   14 rue Juliette Wytsman
   B-1050 BRUXELLES

Responsible organisations

9. Ir. J. Schoofs
PFN d'Infoterra
Services du Premier Ministre
Programmation de la Politique Scientifique
8 rue de la Science
B-1040 BRUXELLES
Tel: (32 2) 2383411
Fax: (32 2) 2305912

10. Dr T. Lakhanisky,
Institut d'Hygiène et d'Epidemiologie
rue Juliette Wytsman 14
B-1050 BRUXELLES
Tel: (32 2) 642 5104
Fax: (32 2) 642 5001

Belize

Responsibility	
Main authority responsible for approval/registration of pesticides	2
Main authority responsible for pesticide use directives, permits and training	-
Main authority responsible for establishing national maximum pesticide residue limits	1
Main authority responsible for control of pesticide residues in food	1
Main laboratory for analysis of pesticide residues in food	1
Main laboratory for quality control of pesticide products	
Main organisation responsible for applied research regarding pesticide use. Information on related regional research	-
Designated national authority for notifications and information regarding banned and severely restricted pesticides including the PIC procedure	2
Codex Alimentarius contact point	3
National IRPTC correspondent	2
Infoterra national focal point	4
National IFCS focal point	-
National association of pesticide manufacturers	5

Responsible organisations

1. Central Veterinary Laboratory
   P.O. Box 181
   Saint Joseph Street
   BELIZE CITY
   Tel: (501) 244 794
   Fax: (501) 230 117

2. The Secretary
   Pesticides Control Board
   Department of Agriculture
   Central Farm
   CAYO DISTRICT
   Tel: (501) 92 2640, 92 3772
   Fax: (501) 92 2640

3. Mr. S G Arthurs
   Bureau of Standards
   P.O. Box 1647
   BELIZE CITY
   Tel: (501) 262 314
   Fax: (501) 822 923

4. Mr. Jaime Villanueva
   Infoterra NFP
   Fisheries Department
   P.O. Box 148
   BELIZE CITY
   Tel: (501 2) 32623, 44552
   Fax: (501 2) 32983

5. James Brodies and Company
   P.O. Box 365
   BELIZE CITY
   Tel: (501 2) 79285
   Fax: (501 2) 75593

Benin
République du Bénin

Responsibility	
Main authority responsible for approval/registration of pesticides	1
Main authority responsible for pesticide use directives, permits and training	
Main authority responsible for establishing national maximum pesticide residue limits	1
Main authority responsible for control of pesticide residues in food	6
Main laboratory for analysis of pesticide residues in food	
Main laboratory for quality control of pesticide products	-
Main organisation responsible for applied research regarding pesticide use. Information on related regional research	1
Designated national authority for notifications and information regarding banned and severely restricted pesticides including the PIC procedure	2
Codex Alimentarius contact point	3
National IRPTC correspondent	-
Infoterra national focal point	4
National IFCS focal point	7
National association of pesticide manufacturers	5

Responsible organisations

1. Service Protection des Vegetaux
   B.P. 58
   PORTO-NOVO
   Tel: 213290
   Fax: 300446

2. Monsieur le Directeur
   Direction de l'Agriculture
   Ministère du Développement rural et de l'action coopérative
   B.P. 03-2900
   COTONOU

3. Secretariat de la Commission Nationale du Codex Alimentarius
   Direction de l'Alimentation et de la Nutrition Appliquée (DANA)
   Ministère du Developpement Rural
   B.P. 295
   PORTO-NOVO
   Tel: 212670

4. Asse A. Sévérin, Directeur
   PFN d'Infoterra
   Direction de l'Aménagement du Territoire
   Ministère de l'Environnement, de l'Habitat et de l'Urbanisme
   B.P. 01-3621
   COTONOU
   Tel: (229) 300742, 300541
   Telex: 5118 miniplan

5. Groupement Professionel des Distributeurs d'Intrants Agricoles (GPDIA)
   s/c de Monsieur Balle
   Janvier Omar Samac/CIBA
   B.P. 8123
   COTONOU
   Tel: 314770

6. Direction de l'Alimentation et de la Nutrition Appliquée
   B.P. 295
   PORTO-NOVO

7. Mr. Lucien K Owoladi
   Ingenieur Forestier
   Ministère de l'Environnement, de l'Habitat et de l'Urbanisme
   B.P. 01-3621
   COTONOU
   Tel: (229 31) 5596
   Fax: (229 31) 5081

Bhutan
Druk - Yul

Responsibility	
Main authority responsible for approval/registration of pesticides	1, 2
main authority for pesticide use directives, permits and training	
Main authority responsible for establishing national maximum pesticide residue limits	2
Main authority responsible for control of pesticide residues in food	1
Main laboratory for analysis of pesticide residues in food	3
Main laboratory for quality control of pesticide products	
Main organisation responsible for applied research regarding pesticide use. Information on related regional research	3
Designated national authority for notifications and information regarding banned and severely restricted pesticides including the PIC procedure	4
Codex Alimentarius contact point	3
National IRPTC correspondent	-
Infoterra national focal point	5
National IFCS focal point	-
National association of pesticide manufacturers	-

Responsible organisations

1. Ministry of Agriculture
   THIMPU
   Tel: 22129, 22168
   Fax: (795 2) 23153

2. National Environment
   Commission (NEC)
   Royal Government of Bhutan
   THIMPU
   Tel: 23384
   Fax: (975) 23385

3. National Plant Protection
   Centre (NPPC)
   Department of Agriculture
   SIMTOKHA
   Tel: 23092
   Fax: (975) 23940

4. The Secretary
   Ministry of Agriculture
   THIMPHU
   Tel: 22129, 22168

5. Infoterra NFP
   Communication Officer
   National Environment
   Commission
   P.O. Box 466
   THIMPHU
   Tel: (975 2) 23384
   Fax: (975 2) 23385

Bolivia
República de Bolivia

Responsibility	
Main authority responsible for approval/registration of pesticides	1
Main authority responsible for pesticide use directives, permits and training	
Main authority responsible for establishing national maximum pesticide residue limits	1
Main authority responsible for control of pesticide residues in food	1, 2, 3, 4
Main laboratory for analysis of pesticide residues in food	1, 5
Main laboratory for quality control of pesticide products	
Main organisation responsible for applied research regarding pesticide use. Information on related regional research	1
Designated national authority for notifications and information regarding banned and severely restricted pesticides including the PIC procedure	6, 10
Codex Alimentarius contact point	7
National IRPTC correspondent	8
Infoterra national focal point	9
National IFCS focal point	-
National association of pesticide manufacturers	-

Responsible organisations

1. Laboratorio de Análisis de Calidad y Reciduos de Plaguicidas en Alimentos
Ministerio de Asuntos Campecinos y Agropecuarios (MACA)
Avda. Camacho 1471
LA PAZ
Tel: 376 679, 782 366

2. Dirección Nacional de Normas y Tecnología
Ministerio de Industria y Comercio
Avda. Camacho esq. Bueno
LA PAZ
Tel: 372 047

3. Laboratorio de Bioquímica Nutricional (INLASA)
Calle Nestor Morales Villazón
LA PAZ
Tel: 376 670

4. Oficina de Control de Calidad de Alimentos y Bebidas
Ministerio de Previsión Social y Salud Pública
LA PAZ
Tel: 375 474

5. Division de Nutrición
Ministerio de Previsión Social y Salud Pública
Avda. Santa Cruz
Edificio Lotería 8 vo. piso
LA PAZ
Tel: 375 478

6. The Director
Control de Calidad Ambiental
Subsecretaria de Medio Ambiente
Secretaria Nacional de Recursos Naturales y Medio Ambiente
Ministerio de Desarollo Sostenible y Medio Ambiente
Av. Arce No. 2147
Casilla No. 12814
LA PAZ
Tel: 372 063, 367 449
Fax: 591(2) 392 892

7. Dr. Edgar Caceres
Director, Instituto Nacional de Alimentación y Nutrición
Calle Ballivian No. 1421
Casilla 4752
LA PAZ
Tel: 37891
Fax: 356 471

8. El Director
Instituto de Investigaciones Quimicas
FCPN - UMSA
Universidad Boliviana
Casilla 303
LA PAZ

9. Dr. Carlos Aguirre Bastos
CNC de Infoterra
Academia Nacional de Ciencias de Bolivia
Av. 16 de Julio No. 1732
Casilla 5829
LA PAZ
Tel: (591 2) 363 990
Fax: (591 2) 379 681

10. Direccion Nacional de
    Produccion y Proteccion
    Agricola
    Subsecretaria Nacional de
    Agricultura y Ganaderia
    Ministerio de Hacienda y
    Desarollo Economico
    Av. Camacho No. 1471, 5 piso
    LA PAZ
    Fax: (591 2) 357 535

Botswana
Republic of Botswana

Responsibility	
Main authority responsible for approval/registration of pesticides	1, 2
Main authority for pesticide use directives, permits and training	
Main authority responsible for establishing national maximum pesticide residue limits	3
Main authority responsible for control of pesticide residues in food	2, 3
Main laboratory for analysis of pesticide residues in food	2, 3
Main laboratory for quality control of pesticide products	
Main organisation responsible for applied research regarding pesticide use. Information on related regional research	4
Designated national authority for notifications and information regarding banned and severely restricted pesticides including the PIC procedure	-
Codex Alimentarius contact point	5
National IRPTC correspondent	-
Infoterra national focal point	6
National IFCS focal point	7
National association of pesticide manufacturers	-

Responsible organisations

1. Ministry of Commerce and Industries
GABORONE

2. Ministry of Agriculture and Livestock
GABORONE

3. Ministry of Health
P.O. Box 993
GABORONE

4. Department of Agricultural Research
Private Bag 0033
GABORONE
Tel:52381

5. The Head
National Food Control Laboratory
Ministry of Health
Private Bag 00269
GABORONE

6. Ms Mmapula Matsheka
Infoterra National Focal Point
Deparment of Town and Regional Planning
Ministry of Local Government and Lands
Private Bag 0042
GABORONE
Tel: (267) 354 277, 354 285
Fax: (267) 313 280, 354 273

7. Chief Community Health Officer
Community Health Services Division
Private Bag 00269
GABORONE
Tel: (267) 374 351, 374 355
Fax: (267) 374 354

Brazil

República Federativa do Brasil

Responsibility	
Main authority responsible for approval/registration of pesticides	1, 2, 8
Main authority for pesticide use directives, permits and training	2
Main authority responsible for establishing national maximum pesticide residue limits	2
Main authority responsible for control of pesticide residues in food	1, 2, 3, 4, 5
Main laboratory for analysis of pesticide residues in food	2, 3, 4, 5
Main laboratory for quality control of pesticide products	
Main organisation responsible for applied research regarding pesticide use. Information on related regional research	6, 7
Designated national authority for notifications and information regarding banned and severely restricted pesticides including the PIC procedure	8
Codex Alimentarius contact point	9
National IRPTC correspondent	8
Infoterra national focal point	10
National IFCS focal point	12
National association of pesticide manufacturers	11

Note

Poisons information/control centres in Belem, Belo Horizonte, Botucatu, Brasilia, Campinas, Campo Grande, Cuiaba, Curitiba, Florianopolis, Fortaleza, Goiana, Jabaquara, Joao Pessoa, Londrina, Manuas, Marilia, Maringa, Natal, Niteroi, Porto Alegre, Porto Velho, Presidente Prudente, Recife, Ribeirao Preto, Salvador, Santos, Sao Jose Dos Campos, Sao Jose do Rio Preto, Sao Luis, Sao Paulo, Sorocoba, Taubate, Uberlandia and Vitoria.

Responsible organisations

1. Ministério da Agricultura
Esplanada dos Ministérios
Bl.D
70000 BRASILIA DF
Tel: (61) 226 9575
Telex: 1930

2. Ministério da Saúde
Secretaria Nacional de Vigilancia Sanitária (SNVS)
Esplanada dos Ministérios
Bl.11
70000 BRASILIA DF
Tel: (61) 226 9961, 226 9903
Fax: (61) 225 6056

3. Ministério da Agricultura
Laboratório Nacional de Referéncia Vegetal (LANARV)
Esplanada dos Ministérios
Bl.D
70000 BRASILIA DF
Tel:(61) 226 9799

4. Ministério da Saúde
Fundacáo Osvaldo Cruz
Instituto Nacional de Controle de Qualidade e Saúde (INCQS/ FIOCRUZ)
Manguinhos-Rio de Janeiro
21050 RIO DE JANEIRO
Tel: (21) 290 0392

5. Ministério da Agricultura
Laboratório Nacional de Referéncia Animal (LANARA)
Esplanada dos Ministérios
Bl.D
70000 BRASILIA DF
Tel: (61) 226 9791
Telex 1930

6. Centro Nacional de Defensivos Agrícolas
Av. Francisco Glicério 957
13100 CAMPINAS SP

7. Empresa Brasileira de Pesquisa Agropecuária (EMBRAPA)
Ed. Venáncio 2.000-B
70000 BRASILIA DF
Tel: (61) 226 4417
Telex: 1620

8. The Director,
Instituto Brasileiro do Meio Ambiente e dos Recursos Naturais Renovaveis (IBAMA)
Sain Av. L-4 Norte, Ed. Sede do IBAMA
70.800-200 BRASILIA DF
Tel: (55 61) 226 7101, 226 6803
Fax: (55 61) 224 5206

9. Divisáo de Organismos Internacionalis Especializados (DIE)
Ministério das Relacoes Exteriores
Pal. do Itamaraty, Anexo I-Sala 418
Explanada dos Ministérios
70170 BRASILIA DF
Tel: (061) 211 6328
Fax: (061) 223 7362

10. Dra. Rita de Cássia do Vale Caribé
CNC de Infoterra
Centro Nacional de Informacáo Ambiental (CNIA)
Instituto Brasileiro do Meio Ambiente e dos Recursos Naturais Renováveis (IBAMA)
SAIN AV. L4 Norte - Ed. Sede
70.800 BRASILIA DF
Tel: (55 61) 226 5094, 1231 2324
Fax: (55 61) 224 5206

11. Associacao Nacional de Defesa Vegetal (ANDEF)
Rua Capitao Antonio Rosa 376, 13° Andar
Jardim Paulistano
01443 SAO PAULO-SP
Tel: (55 11) 881 5033
Fax: (55 11) 853 2637

12. Ms M. Cerqueira
Biologist
Department of Environmental Management
Ministry of the Environment, Water Resources, and the Legal Amazon
Esplanada Dos Ministèros Bloco B, Sala 838
BRASILIA DF
Tel: (5561) 317 1215/317 1244
Fax: (5561) 317 1352

Bulgaria
Republika Bulgaria

Responsibility	
Main authority responsible for approval/registration of pesticides	1, 2, 3, 4, 7
Main authority responsible for pesticide use directives, permits and training	
Main authority responsible for establishing national maximum pesticide residue limits	1, 2, 3, 7
Main authority responsible for control of pesticide residues in food	1, 3
Main laboratory for analysis of pesticide residues in food	3, 5, 6, 11
Main laboratory for quality control of pesticide products	
Main organisation responsible for applied research regarding pesticide use. Information on related regional research	2, 5, 6, 7
Designated national authority for notifications and information regarding banned and severely restricted pesticides including the PIC procedure	7
Codex Alimentarius contact point	8
National IRPTC correspondent	9
Infoterra national focal point	10
National IFCS focal point	-
National association of pesticide manufacturers	-

Note

Poisons information/control centre: National Toxicology Centre, Research Institute of Emergency Medicine N. Pirogov, Section of Clinical Toxicology, Bd. Totlebene 21, Sofia 1606, Tel: (359) 5153346, Fax: (359) 5153234 [Head: Pr. N. Mirtchev]

Responsible organisations

1. Ministerstvo na Narodnoto Zdrave
5 Plostad Lenin
SOFIA
Tel: 8631
Telex: 22430 MNZ

2. Ministry of Agriculture
55, Hr. Botev Blvd
1040 SOFIA
Tel. 8531
Telex: 22325, 22326, 22798

3. Institut på Gastroenterologija i Hranene
Laboratrija po Hranitelna Toksikologija
15 Boul. Dim. Nestorov
1431 SOFIA
Tel: 58101
Telex: 22712 MAPREZ

4. Institut po Higiena i Professionalni Zaboljavanija Sekcija Industrialna Toksikologija
15 Boul. Dim. Nestorov
1431 SOFIA
Tel: 58101
Telex: 22712 MAPREZ

5. Vis Selskostopanski Institut "V. Kolarov"
Katedra Rastitelna Zastita
12 D. Mendeleev
PLOVDIV

6. Institut za Zastita na Rastenijata Kostinbord
Sofiya District
SOFIA

7. National Service for Plant Protection, Quarantine and Agrochemistry
Ministry of Agriculture
55 Bd. Hristo Botev
SOFIA

8. Mr. le Chef de la Section de la Commission du Codex Alimentarius
Union nationale agro-industrielle
55 Boul. Hristo Botev
1000 SOFIA

9. Prof F Kaloyanova
National Centre of Hygiene and Medical Ecology
15 Boul. Dim. Nestorov
1431 SOFIA
Tel: (359 2) 598076 or 5812590
Fax: (359 2) 598148

10. Mrs. Donna Schalitcheva
Infoterra NFP
National Centre for Environment and Sustainable Development
Department of Training and Information
Ministry of Environment
7, Industrialna Str.
1202 SOFIA
Tel: (359 2) 394725, 392027
Fax: (359 2) 392196, 392027

11. Central laboratory for control of residues of pesticides, nitrates and heavy metals
"202" Street No. 2
1220 SOFIA

Burkina Faso

Responsibility

Main authority responsible for approval/registration of pesticides	-
Main authority responsible for pesticide use directives, permits and training	
Main authority responsible for establishing national maximum pesticide residue limits	-
Main authority responsible for control of pesticide residues in food	-
Main laboratory for analysis of pesticide residues in food	-
Main laboratory for quality control of pesticide products	
Main organisation responsible for applied research regarding pesticide use. Information on related regional research	-
Designated national authority for notifications and information regarding banned and severely restricted pesticides including the PIC procedure	1
Codex Alimentarius contact point	2
National IRPTC correspondent	-
Infoterra national focal point	3
National IFCS focal point	-
National association of pesticide manufacturers	-

Responsible organisations

1. Monsieur le Directeur
   Service des analyses et application de la législation phytosanitaire
   Direction de la protection des vegetaux et du conditionnement
   OUAGADOUGOU

2. Ministère du Développement Rural
   B.P. 7010
   OUAGADOUGOU

3. M. Bruno R. Salo
   PFN d'Infoterra
   Direction de la Prevention des Pollutions et de l'Amelioration due Cadre de Vie (DIPAC)
   Ministère de l'Environnement et de l'Eau
   B.P. 7044
   OUAGADOUGOU 03
   Tel: (226) 311919
   Fax: (226) 306767, 301232

Burundi

République du Burundi, Republika y'u Burundi

Responsibility	
Main authority responsible for approval/registration of pesticides	
Main authority for pesticide use directives, permits and training	
Main authority responsible for establishing national maximum pesticide residue limits	
Main authority responsible for control of pesticide residues in food	
Main laboratory for analysis of pesticide residues in food	
Main laboratory for quality control of pesticide products	
Main organisation responsible for applied research regarding pesticide use. Information on related regional research	
Designated national authority for notifications and information regarding banned and severely restricted pesticides including the PIC procedure	1
Codex Alimentarius contact point	3
National IRPTC correspondent	2
Infoterra national focal point	4
Natiomal IFCS focal point	-
National association of pesticide manufacturers	-

Responsible organisations

1. Monsieur le Directeur
   Departement de la défense des cultures
   B.P.114
   GITEGA

2. Monsieur le Directeur général de l'Urbanisme et de l'Habitat
   B.P. 1860
   BUJUMBURA

3. Bureau Burundais de normalisation et controle de la qualite (BNN)
   B.P. 3535
   BUJUMBURA
   Tel: 22 1815, 22 1577

4. Dr. Gaspard Bikwemu
   PFN d'Infoterra
   Institut national pour l'Environnement et la Conservation de la Nature (INECN)
   B.P. 56
   GITEGA
   Tel: (257 40) 2071
   Fax: (257 40) 2075

Cambodia

Preah Reach Ana Pak Kampuchea

Responsibility	
Main authority responsible for approval/registration of pesticides	
Main authority responsible for pesticide use directives, permits and training	
Main authority responsible for establishing national maximum pesticide residue limits	
Main authority responsible for control of pesticide residues in food	
Main laboratory for analysis of pesticide residues in food	
Main laboratory for quality control of pesticide products	
Main organisation responsible for applied research regarding pesticide use. Information on related regional research	
Designated national authority for notifications and information regarding banned and severely restricted pesticides including the PIC procedure	-
Codex Alimentarius contact point	1
National IRPTC correspondent	-
Infoterra national focal point	2
National IFCS focal point	-
National association of pesticide manufacturers	

Responsible organisations

1. Mr. Lim Thearith
   Assistant Quality
   Control Service
   KAMCONTROL
   50 E/144 Street
   PHNOM PENH
   Tel: 2 3485
   Fax: (855) 23 26166

2. Infoterra NFP
   Ministry for Environment
   48, Samdech Preah Sihanouk
   Tonle Bassac, Chamkamon
   PHNOM PENH
   Tel: (855) 232 7844
   Fax: (855) 232 7844

Cameroon
République du Cameroun, Republic of Cameroon

Responsibility	
Main authority responsible for approval/registration of pesticides	-
Main authority responsible for pesticide use directives, permits and training	
Main authority responsible for establishing national maximum pesticide residue limits	-
Main authority responsible for control of pesticide residues in food	-
Main laboratory for analysis of pesticide residues in food	-
Main laboratory for quality control of pesticide products	
Main organisation responsible for applied research regarding pesticide use. Information on related regional research	1
Designated national authority for notifications and information regarding banned and severely restricted pesticides including the PIC procedure	2
Codex Alimentarius contact point	3
National IRPTC correspondent	4
Infoterra national focal point	5
National IFCS focal point	-
National association of pesticide manufacturers	6

Responsible organisations

1. Institute de la Recherche Agronomique (IRA)
   B.P. 2123
   YAOUNDÉ
   Tel: 232644
   Telex: 1140

2. Monsieur le Directeur
   Direction de l'agriculture
   Sous-Direction de la protection des végétaux
   Ministère de l'agriculture
   YAOUNDÉ

3. Ministère du Plan et de l'Aménagement du Territoire
   B.P. 1004
   YAOUNDÉ

4. M. le Directeur Général
   Direction de la Recherche Scientifique et Technique
   Ministère de l'Enseignement Supérieur et de la Recherche Scientifique
   B.P. 1457
   YAOUNDÉ

5. M. Sapock Ndem Samuel
   PFN d'Infoterra
   Directeur de l'Aménagement du Territoire et de l'Environnement
   Ministére du Plan et de l'Aménagement du Territoire
   YAOUNDÉ
   Tel: (237) 233672

6. U.P.A.C.
   c/o Rhône-Poulenc Afrique Centrale
   B.P. 4007
   DOUALA
   Tel: (237)433179
   Fax: (237)433178

Canada

Responsibility	
Main authority responsible for approval/registration of pesticides*	1, 11
Main authority responsible for pesticide use directives, permits and training*	
Main authority responsible for establishing national maximum pesticide residue limits*	2
Main authority responsible for control of pesticide residues in food*	2,3
Main laboratory for analysis of pesticide residues in food*	2,4
Main laboratory for quality control of pesticide products*	
Main organisation responsible for applied research regarding pesticide use. Information on related regional research*	5
Designated national authority for notifications and information regarding banned and severely restricted pesticides including the PIC procedure*	6
Codex Alimentarius contact point	7
National IRPTC correspondent	8
Infoterra national focal point	9
National IFCS focal point	12
National association of pesticide manufacturers	10

Note

Poisons information/control centres in Calgary, Ottawa, Quebec, Regina, Saskatoon, Toronto and Winnipeg

* Canada is in the process of determining how these functions will be performed under the new Pest Management Regulatory Authority

Responsible organisations

1. Plant Industry Directorate
Food Production and Inspection Branch
Agriculture and Agri-Food Canada
OTTAWA, Ontario K1A 0C5
Tel: (613) 993 4544
Fax: (613) 998 1312

2. Food Directorate
Health Protection Branch
Health Canada
OTTAWA, Ontario K1A 0L2
Tel: (613) 957 1821
Fax: (613) 957 1784

3. Field Operations Directorate
Health Protection Branch
Health Canada
OTTAWA, Ontario K1A 0L2
Tel: (613) 957 1794
Fax: (613) 941 0400 or 954-8637

4. Pesticide Laboratory
Laboratory Services Division
Agriculture and Agri-Food Canada
OTTAWA, Ontario K1A 0C5
Tel: (613) 995 4907
Fax: (613) 992 5819

5. Scientific Information Retrieval Section
Research Branch
Agriculture and Agri-Food Canada
OTTAWA, Ontario K1A 0C5
Tel: (613) 995 7084
Fax: (613) 992 7909

6. The Director
Information Division
Plant Industry Directorate
Agriculture and Agri-Food Canada
OTTAWA, Ontario K1A 0C5
Tel: (613) 993 4544
Fax: (613) 998 1312

7. Mr. Ron B. Burke
Bureau of Food Regulatory, International and Interagency Affairs
Food Directorate
Health Protection Branch
Health Canada
Tunney's Pasture
Room 200,
H.P.B. Building
OTTAWA, Ontario K1A 0L2
Tel: (613) 957 1748
Fax: (613) 941 3537

8. The Director,
Bureau of Chemical Hazards
Environmental Health Directorate
Health Protection Branch
Health Canada
Attn: Ms. Véronique Morisset
Environmental Health Centre
Tunney's Pasture
OTTAWA, Ontario K1A 0L2
Tel: (613) 957 1506
Fax: (613) 952 9798

Responsible organisations

9. Ms. Monica Czanyo
Infoterra NFP
Departmental Library
Environment Canada
2nd floor,
Place Vincent Massey
351 St. Joseph Blvd.
HULL, Quebec K1A 0H3
Tel: (819) 997 2575
Fax: (819) 997 5349

10. Crop Protection Institute of Canada (CPIC)
21 Four Seasons Place
Suite 627
ETOBICOKE, Ontario M9B 6J8
Tel: (416) 622 9771
Fax: (416) 622 6764

11. Pesticide Division
Commercial Chemicals Evaluation Branch
Environmental Protection Service
Environment Canada
OTTAWA, Ontario
Tel: (819) 997 3450
Fax: (819) 994 3930

12. Dr. J. R. Hickman
Director General,
Environmental Health Directorate
Health Canada
Tunney's Pasture
OTTAWA, Ontario K1A OL2
Tel: (613) 954 0291
Fax: (613) 952 9798

Cape Verde
República de Cabo Verde

Responsibility	
Main authority responsible for approval/registration of pesticides	1
Main authority responsible for pesticide use directives, permits and training	
Main authority responsible for establishing national maximum pesticide residue limits	2
Main authority responsible for control of pesticide residues in food	2
Main laboratory for analysis of pesticide residues in food	2
Main laboratory for quality control of pesticide products	
Main organisation responsible for applied research regarding pesticide use. Information on related regional research	2
Designated national authority for notifications and information regarding banned and severely restricted pesticides including the PIC procedure	2
Codex Alimentarius contact point	3
National IRPTC correspondent	-
Infoterra national focal point	4
National IFCS focal point	-
National association of pesticide manufacturers	-

Responsible organisations

1. Ministerio das Pescas, Agricultura e Animaço Rural (MPAAR)
   Direcco Geral do Fomento Agrario
   PRAIA

2. Instituto Nacional de Investigaço e Desenvolvimento Agrário (INIDA)
   Departmento de Agricultura
   Division de la Protection des Vegetaux
   C.P. 50
   PRAIA
   Tel. (238) 611363
   Fax: (238) 614054

3. Ministerio das Pescas, Agricultura y Animaço Rural (MPAAR)
   Gabinete de Estudo e Planeamento
   C.P.115
   PRAIA

4. M. Daniel Avelino Pires
   PFN d'Infoterra
   Centre de Documentation et Information pour le Developpement (CDID)
   B.P. 120
   PRAIA
   Tel: (238) 613969
   Fax: (238) 611527

Central African Republic
République Centrafricaine

Responsibility

Main authority responsible for approval/registration of pesticides	
Main authority for pesticide use directives, permits and training	
Main authority responsible for establishing national maximum pesticide residue limits	
Main authority responsible for control of pesticide residues in food	
Main laboratory for analysis of pesticide residues in food	
Main laboratory for quality control of pesticide products	
Main organisation responsible for applied research regarding pesticide use. Information on related regional research	
Designated national authority for notifications and information regarding banned and severely restricted pesticides including the PIC procedure	1
Codex Alimentarius contact point	2
National IRPTC correspondent	-
Infoterra national focal point	3
National IFCS focal point	-
National association of pesticide manufacturers	

Responsible organisations

1. Monsieur le Directeur
   Protection des vegetaux
   Ministère du développement rural
   BANGUI

2. Ministere des Eaux, des Forêts, de la Chasse et de la Peche
   BANGUI

3. M. Emarou Clement
   PFN d'Infoterra
   Chef de Service de l'Hygiene du Milieu
   Direction de l'Assainissement et de la Salubrite de l'Environnement
   B.P. 711
   BANGUI

Chad

Jumhuriyah Tshad, République du Tchad

Responsibility	
Main authority responsible for approval/registration of pesticides	
Main authority responsible for pesticide use directives, permits and training	
Main authority responsible for establishing national maximum pesticide residue limits	
Main authority responsible for control of pesticide residues in food	
Main laboratory for analysis of pesticide residues in food	
Main laboratory for quality control of pesticide products	
Main organisation responsible for applied research regarding pesticide use. Information on related regional research	
Designated national authority for notifications and information regarding banned and severely restricted pesticides including the PIC procedure	1
Codex Alimentarius contact point	2
National IRPTC correspondent	-
Infoterra national focal point	3
National IFCS focal point	-
National association of pesticide manufacturers	

Responsible organisations

1. Direction de Protection des Végétaux et du Conditionnement (DPVC)
   Ministere de l'agriculture
   B.P. 441
   N'DJAMÉNA
   Tel: 513692
   Fax: 512219

2. Direction du génie sanitaire et de l'assainissement
   Sous-direction de l'assainissement
   B.P. 440
   N'DJAMÉNA

3. PFN d'Infoterra
   Secretariat d'etat charge de l'environnement
   Direction des Forêts et de la Protection de l'Environnement
   B.P. 447
   N'DJAMÉNA
   Tel: (235) 2121, Ext 344, 417

Chile
República de Chile

Responsibility	
Main authority responsible for approval/registration of pesticides	1
Main authority responsible for pesticide use directives, permits and training	-
Main authority responsible for establishing national maximum pesticide residue limits	2
Main authority responsible for control of pesticide residues in food	5
Main laboratory for analysis of pesticide residues in food	3
Main laboratory for quality control of pesticide products	1
Main organisation responsible for applied research regarding pesticide use. Information on related regional research	4, 6
Designated national authority for notifications and information regarding banned and severely restricted pesticides including the PIC procedure	1, 5
Codex Alimentarius contact point	2
National IRPTC correspondent	5
Infoterra national focal point	8
National IFCS focal point	7
National association of pesticide manufacturers	9

Responsible organisations

1. Ministerio de Agricultura
Servicio Agrícola y Ganadero
Departamento de Protección Agrícola
Av. Bulnes 140, Piso 8
SANTIAGO
Tel: 671 2323, 271 5131
Fax: 672 1812

2. Ministerio de Salud Publica
Monjitas 689, 5 piso
SANTIAGO
Tel: 639 1198
Fax: 638 4377

3. Instituto de Salud Pública de Chile
Maraton 1000
Mail Box 48
SANTIAGO
Tel: 239 1105
Fax: 238 4536

4. University of Chile
Faculty of Agricultural Sciences & Forestry
Department of Plant Protection
Santa Rosa No. 11315
Mail Box 1004
SANTIAGO
Tel: 541 7703
Fax: 541 7955

5. Ministerio de Salud Publica
Depatamento de Programas Sobre de Ambiente
Mac-Iver No. 541
SANTIAGO

6. Ministerio de Agricultura
Instituto de Investigaciones Agropecuarias (INIA)
Fidel Oteiza No. 1956, pisos 11,12
Providencia
SANTIAGO
Tel: 225 2118
Fax: 225 8773

7. Sr J. Monréal Urrutia
Ministerio de Salud
Monjitas No. 514
SANTIAGO
Tel: (639) 4001
Fax: (639) 7110

8. Ms Ana Maria Prat
CNC de Infoterra
Departamento de Informacion
National Commission of Scientific and Technological Research - CONICYT
Casilla 297 V
SANTIAGO 21
Tel: (56 2) 274 4537
Fax: (56 2) 209 6729

9. Asociación Nacional de Fabricantes e Importadores de Plaguicidas Agricolas (AFIPA)
Providencia 2601
Oficina 11
SANTIAGO
Tel: (56 2) 231 7406
Fax: (56 2) 231 7406

China

Chung-hua Jen-min Kung-ho-kuo, Zhonghua Renmin Gongheguo

Responsibility	
Main authority responsible for approval/registration of pesticides	1
Main authority responsible for pesticide use directives, permits and training	
Main authority responsible for establishing national maximum pesticide residue limits	2
Main authority responsible for control of pesticide residues in food.	2, 3
Main laboratory for analysis of pesticide residues in food	2, 3
Main laboratory for quality control of pesticide products	
Main organisation responsible for applied research regarding pesticide use. Information on related regional research	1
Designated national authority for notifications and information regarding banned and severely restricted pesticides including the PIC procedure	4, 5
Codex Alimentarius contact point	6
National IRPTC correspondent	4, 9
Infoterra national focal point	7
National IFCS focal point	8
National association of pesticide manufacturers	-

Note

Poisons information/control centre: Centre for Acute Poisoning Control, Institute of Occupational Medicine, Chinese Academy of Preventive Medicine, 29 Nan Wei Road, Beijing 100050, Tel: (86) 301 6891, Fax: (86) 1 301 4323 [Head: Dr Zhang Shoulin]

Responsible organisations

1. Ministry of Agriculture, Animal Husbandry and Fisheries
Huopingli
BEIJING

2. Ministry of Public Health
Houhai
BEIJING

3. State Administration for the Inspection of Import and Export Commodities of the Peoples Republic of China
12 Jiangoumenwai
BEIJING
Tel: 594761, 553861
Telex: 210076 saci

4. National Environmental Protection Agency (NEPA)
Attn: Mrs Sun Lijin
115, Xizhimennei
Nanxiaojie
BEIJING 100035
Tel: 6165635 Ext. 5812
Fax: 6151762

5. The Director
Institute for the Control of Agrochemicals
Ministry of Agriculture
BEIJING
Liang Ma Quiao,
Chaoyang
BEIJING
Tel: (86 10) 502 5929
Fax: (86 10) 502 5929

6. Mr Xu Guanghua
Department of Science and Technology
Ministry of Agriculture, Animal Husbandry and Fisheries
BEIJING

7. Prof. Kangsheng Zhang
Infoterra NFP
Research Centre for Eco-Environmental Sciences
Academia Sinica
P.O. Box 2871
BEIJING 100085
Tel: (86 10) 2545511, Ext. 2139
Fax: (86 10) 2555381

8. Mr Yuxiang Zang
Deputy Director General
Department of Pollution Control
National Environmental Protection Agency
115, Zizhimennei Nanxiaojie
BEIJING 100035
Tel: (86 10) 6615 53366
Fax: (86 10) 6615 1762

9. The Director of IRPTC/China
Institute of Environmental Health Monitoring
Chinese Academy of Preventive Medicine
Attn: Dr Xu Li-Juan
7 Pan Jia Yuan
Chao Yang District
BEIJING 100021

Colombia
República de Colombia

Responsibility	
Main authority responsible for approval/registration of pesticides	1, 2, 3, 4, 6
Main authority responsible for pesticide use directives, permits and training	
Main authority responsible for establishing national maximum pesticide residue limits	1, 2, 3, 5
Main authority responsible for control of pesticide residues in food	1, 2, 4, 5, 7
Main laboratory for analysis of pesticide residues in food	2, 5, 8
Main laboratory for quality control of pesticide products	
Main organisation responsible for applied research regarding pesticide use. Information on related regional research	2, 9
Designated national authority for notifications and information regarding banned and severely restricted pesticides including the PIC procedure	2, 10, 11
Codex Alimentarius contact point	12
National IRPTC correspondent	16
Infoterra national focal point	13
National IFCS focal point	15
National association of pesticide manufacturers	14

Note

Poisons information/control centres in Bogota and Medellin

Responsible organisations

1. Ministerio de Salud
   Cra 13, No 32-71, Piso 14
   BOGOTA

2. Instituto Colombiano
   Agropecuario (ICA)
   Calle 37 No. 8-43, Piso 4 y 5
   Apartado aereo 6984
   BOGOTA
   Tel: (57 1) 285 5520
   Fax: (57 1) 285 4351

3. Ministerio de Salud
   Sección de Toxicología
   Attn: Dr. Pena Martinez
   Avda Caracas No 32-15,
   Oficina 201
   BOGOTA D.E.
   Tel: 245 9228,
   287 4589

4. Instituto Colombiano
   Agropecuario (ICA)
   Attn: P C Navarro
   Director, Division Insumos
   Agricolas
   Calle 37 No. 8-43, Piso 5
   BOGOTA

5. Instituto Nacional de los
   Recursos Naturales
   Renovables y del Ambiente
   (INDERENA)
   Carrera 109, No 20-30
   BOGOTA
   Tel: 232 9716

6. Ministerio del Ambiente
   Calle 16, No 6-66, Piso 30
   BOGOTA

7. Instituto Nacionale de Salud
   Avenida El Dorado
   Carrera 50
   BOGOTA

8. Instituto Colombiano
   Agropecuario (ICA)
   Laboratorio Nacional de
   Insumos Agricolas
   Carlos A. Salcedo S.
   Kilometro 14 Via Mosquera
   Apartado Aereo
   151123 BOGOTA

9. Unidad de Evaluación y
   Análisis del Riesgos
   INDERENA
   P.O. Box A.A. 29706
   BOGOTA

10. Jefe,
    Division Sustancias
    Potencialmente Toxicas
    Ministerio de Salud
    Calle 55 No. 10-32
    Bloque B piso 3
    BOGOTA
    Tel: (57 1) 245 9228
    Fax: (57 1) 282 0003

11. Director General
    Instituto Nacional de los
    Recursos Naturales
    Renovables y del Ambiente
    Ministerio de Agricultura
    Carrera 10 No. 20-30,
    Of. 204
    Apartado aereo 13458
    BOGOTA
    Tel: (57 1) 284 2427
    Fax: (57 1) 285 9987

12. Subdireccion de Control de Factores de Riesgo del Consumo
Ministerio de Salud
Calle 55 No. 10-32,
Bloque B
BOGOTA D.E.
Tel: 217 9900
Fax: 235 8577

13. Ms. Hilda Dugand Carling
Jefe, Sección Información y Documentación
Coordinador Nacional Infoterra
SEDE Carrera 10 NRO 2030 Santafe
BOGOTA D.C.
Telex: 44428 inde co

14. Asociacion Nacional de Industriales (ANDI)
Carrera 13 No. 26-45, piso 6
BOGOTA
Tel: (57 1) 334 9620
Fax: (57 1) 291 3168, 281 3188

15. Mr J A Lopez
Oficina de Cooperacion Internacional
Ministerio del Medio Ambiente
SANTA FE DE BOGOTA
Tel: (571) 342 7629
Fax: (571) 336 0968

16. Jefe,
Division de Sustancias
Potencialmente Toxicas
Ministerio de Salud
Calle 55 no 10-32,
Bloque B-Piso 30
BOGOTA, D.E.

Comoros

Jumhuriyat al-Qumur al-Ittihadiyah al-Islamiyah, République Fédérale Islamique des Comores

Responsibility	
Main authority responsible for approval/registration of pesticides	
Main authority responsible for pesticide use directives, permits and training	
Main authority responsible for establishing national maximum pesticide residue limits	
Main authority responsible for control of pesticide residues in food	
Main laboratory for analysis of pesticide residues in food	
Main laboratory for quality control of pesticide products	
Main organisation responsible for applied research regarding pesticide use. Information on related regional research	
Designated national authority for notifications and information regarding banned and severely restricted pesticides including the PIC procedure	1
Codex Alimentarius contact point	-
National IRPTC correspondent	-
Infoterra national focal point	2
National IFCS focal point	-
National association of pesticide manufacturers	

Responsible organisations

1. Institut national de recherche pour le developpement de l'agriculture, de la peche et de l'environnement
   B.P. 289
   MORONI
   Tel: (269) 730 953
   Fax: (269) 732 511

2. M. Aboulhouda Youssouf Ibrahim
   PFN d'Infoterra
   Le Directeur Général de l'Environnement
   Direction Générale de l'Environnement
   B.P. 41
   MORONI
   Tel: (269) 74 46 36
   Fax: (269) 74 46 32

Congo

République du Congo

Responsibility	
Main authority responsible for approval/registration of pesticides	1, 2
Main authority responsible for pesticide use directives, permits and training	
Main authority responsible for establishing national maximum pesticide residue limits	-
Main authority responsible for control of pesticide residues in food	-
Main laboratory for analysis of pesticide residues in food	-
Main laboratory for quality control of pesticide products	
Main organisation responsible for applied research regarding pesticide use. Information on related regional research	2
Designated national authority for notifications and information regarding banned and severely restricted pesticides including the PIC procedure	3, 6
Codex Alimentarius contact point	4
National IRPTC correspondent	-
Infoterra national focal point	5
National IFCS focal point	-
National association of pesticide manufacturers	-

Responsible organisations

1. Ministère de la Santé et des Affaires Sociales
Direction de l'Hygiène Générale
B.P. 155
BRAZZAVILLE
Tel: 811 852

2. Ministère de l'Agriculture et d'Élevage
Service de la Protection des Végétaux
B.P. 387
BRAZZAVILLE
Tel: 811 813

3. Monsieur le Directeur
Direction de la protection des végétaux
Ministre de l'agriculture, des eaux et forêts, de l'elevage et de l'environnement
B.P. 387
BRAZZAVILLE
Tel: (242) 832 908
Fax: (242) 832 908

4. Représentant de la FAO au Congo et à Sao Tomé et Principe
B.P. 972
BRAZZAVILLE
Tel: (242) 830 346
Fax: (242) 835 502, 837 987

5. Dr. Marius Issanga-Ngamissimi
Point Focal National Infoterra
Direction Générale de l'Environnement
B.P. 958
BRAZZAVILLE
Tel: (242) 833 046
Fax: (242) 837 150

6. Monsieur le Directeur général
Direction générale de l'environnement
B.P. 958
BRAZZAVILLE
Tel: (242) 833 046
Fax: (242) 837 150

Cook Islands

Responsibility	
Main authority responsible for pesticide use directives, permits and training	1
Main authority responsible for establishing national maximum pesticide residue limits	
Main authority responsible for control of pesticide residues in food	1, 2, 3
Main laboratory for analysis of pesticide residues in food	2
Main laboratory for quality control of pesticide products	2,3
Main organisation responsible for applied research regarding pesticide use. Information on related regional research	2,3
Designated national authority for notifications and information regarding banned and severely restricted pesticides including the PIC procedure	4
Codex Alimentarius contact point	1
National IRPTC correspondent	2
Infoterra national focal point	-
National IFCS focal point	-
National association of pesticide manufacturers	-

Responsible organisations

1. Pesticide Board
   Att: Chairman
   P.O. Box 96
   Ministry of Agriculture
   RAROTONGA
   Tel: 28711
   Fax: 21881

2. Chief Health Inspector
   Division of Public Health
   Ministry of Health
   P.O. Box 109
   RAROTONGA
   Tel: 29110
   Fax: 25664

3. Dr. M. Purea
   Director of Research
   P.O. Box 96
   Ministry of Agriculture
   RATOTONGA
   Tel: 28720
   Fax: 21881

4. Secretary of Agriculture
   Ministry of Agriculture
   P.O. Box 96
   Government of the Cook Islands
   RAROTONGA
   Tel: (682) 28711
   Fax: (682) 21881

Costa Rica
República de Costa Rica

Responsibility	
Main authority responsible for approval/registration of pesticides	1
Main authority resåponsible for pesticide use directives, permits and training	
Main authority responsible for establishing national maximum pesticide residue limits	1, 2
Main authority responsible for control of pesticide residues in food	2
Main laboratory for analysis of pesticide residues in food	9
Main laboratory for quality control of pesticide products	
Main organisation responsible for applied research regarding pesticide use. Information on related regional research	1
Designated national authority for notifications and information regarding banned and severely restricted pesticides including the PIC procedure	4
Codex Alimentarius contact point	5
National IRPTC correspondent	6
Infoterra national focal point	7
National IFCS focal point	8
National association of pesticide manufacturers	3

Note

Poisons information/control centre: Centro Nacional de Control de Intoxicaciones, Hospital Nacional de Ninos "Dr Carlos Saenz Herrera", Apartado Postal 1654, San Jose', Tel: (506) 231028, Fax: (506) 554907 [Head: Dr. Damarys Quiros Vega]

Responsible organisations

1. Departamento de Fitopatología
   Dirección Sanidad Vegetal y Comisión Asesora de Plaguicidas
   Apartado 10094
   SAN JOSÉ

2. Departamento de Salud
   Departamento Control Alimentos y Laboratorio Químico
   Apartado 10123
   SAN JOSÉ

3. Cámara de Insumos Agropecuarios
   Apartado 691
   2150 Moravia
   SAN JOSÉ
   Tel: (506) 349819
   Fax: (506) 349932

4. Director General,
   Direccion General de Sanidad Agropecuaria
   Ministerio de Agricultura y Ganaderia
   Apartado Postal 10094
   SAN JOSÉ
   Tel: (506) 231 1939
   Fax: (506) 231 5004

5. Comité Nacional del Codex Alimentarius
   Oficina Nacional de Normas y Unidades de Medida
   Ministerio de Económia, Industria y Comercio
   Apartado 10216
   1000 SAN JOSÉ
   Tel: (506) 220 898
   Fax: (506) 220 898

6. El Director,
   Departamento se Sustancias Toxicas y Medicina del Trabajo
   Ministerio de Salud
   Apartado Postal No. 10 123
   1000 SAN JOSÉ

7. Lic. Max Francisco Cerdas López
   CNC de Infoterra
   Departamento de Información y Documentación
   Consejo Nacional de Investigaciones Científicas y Tecnológicas (CONICIT)
   Apartado 10318
   1000 SAN JOSÉ
   Tel: (506) 244 172
   Fax: (506) 232 576, 349 299

8. Dr. Roberto Castro
   Director del Dpto. de Sustancias Toxicas y Medicina del Trabajo
   Ministerio de Salud
   Apartado 10123
   SAN JOSÉ
   Tel: (506) 233 1081
   Fax: (506) 222 9625

9. Laboratorio de Residuos de Plaguicidas
   Direccion General de Sanidad Vegetal
   Ministerio de Agricultura y Ganaderia
   Apartado Postal 10094
   SAN JOSÉ

Côte d'Ivoire

République de Côte d'Ivoire

Responsibility	
Main authority responsible for approval/registration of pesticides	1
Main authority responsible for pesticide use directives, permits and training	
Main authority responsible for establishing national maximum pesticide residue limits	-
Main authority responsible for control of pesticide residues in food	1, 2, 3
Main laboratory for analysis of pesticide residues in food	4
Main laboratory for quality control of pesticide products	
Main organisation responsible for applied research regarding pesticide use. Information on related regional research	-
Designated national authority for notifications and information regarding banned and severely restricted pesticides including the PIC procedure	5
Codex Alimentarius contact point	6
National IRPTC correspondent	-
Infoterra national focal point	7
National IFCS focal point	-
National association of pesticide manufacturers	8

Responsible organisations

1. Ministère de l'Agriculture
ABIDJAN

2. Ministère de la Production
Animale
ABIDJAN

3. Laboratoire National de la
Santé Publique
Boulevard de Marseille
ABIDJAN

4. Faculté de Pharmacie
d'Abidjan
ABIDJAN

5. Monsieur le Directeur des
Actions Industrielles
B.P. V.50
ABIDJAN
Tel: 226 728, 324 521
Fax: 215 320

6. M. le Secrétaire Général
Comité National pour
l'Alimentation et le
Développement (CNAD)
B.P. V 190
ABIDJAN
Tel: 214934

7. M. Kouakou James Koffi
Manager,
PFN Infoterra
Nour Al Hayat
2e Etage Porte 8
B.P. V 67
ABIDJAN
Tel: (225) 225354
Telex: 23399 mini maci

8. UNIPHYTO
15 B.P. 323
ABIDJAN 15
Tel: (225) 274785
Fax: (225) 271866

Croatia
Republika Hrvatska

Responsibility

Main authority responsible for approval/registration of pesticides	1
Mainauthority responsible for pesticide use directives, permits and training	-
Main authority responsible for establishing national maximum pesticide residue limits	1
Main authority responsible for control of pesticide residues in food	1
Main laboratory for analysis of pesticide residues in food	4
Main laboratory for quality control of pesticide products	-
Main organisation responsible for applied research regarding pesticide use. Information on related regional research	-
Designated national authority for notifications and information regarding banned and severely restricted pesticides including the PIC procedure	1
Codex Alimentarius contact point	2
National IRPTC correspondent	5
Infoterra national focal point	-
National IFCS focal point	3
National association of pesticide manufacturers.	-

Responsible organisations

1. Ministry of Health
   P.O. Box 670
   11000 ZAGREB
   Tel: (385 1) 459 1333
   Fax: (385 1) 459 1341

2. Mrs. Nada Marcovi
   State Office for Standardization and Metrology - DZNM
   Ul. grada Vukovara 78
   10000 ZAGREB
   Tel: (385 1) 613 3444
   Fax: (385 1) 536 688

3. Mr. F Marovic
   Ministry of Labour and Social Welfare
   Prisavlje 14
   10000 ZAGREB
   Tel: (385 1) 616 91111
   Fax: (385 1) 611 4489

4. Croatian National Institute of Public Health
   P.O. Box 684
   10000 ZAGREB
   Tel: (385 1) 276 248
   Fax: (358 1) 276 248

5. State Directorate for Environment
   Ul. grada Vukovara 78
   10000 ZAGREB
   Tel: (385 1) 613 3444
   Fax: (385 1) 537 203

Cuba

República de Cuba

Responsibility	
Main authority responsible for approval/registration of pesticides	1
Main authority responsible for pesticide use directives, permits and training	
Main authority responsible for establishing national maximum pesticide residue limits	1
Main authority responsible for control of pesticide residues in food	2, 3, 4,5
Main laboratory for analysis of pesticide residues in food	3, 4, 5
Main laboratory for quality control of pesticide products	
Main organisation responsible for applied research regarding pesticide use. Information on related regional research	3, 6
Designated national authority for notifications and information regarding banned and severely restricted pesticides including the PIC procedure	7
Codex Alimentarius contact point	8
National IRPTC correspondent	9
Infoterra national focal point	10
National IFCS focal point	-
National association of pesticide manufacturers	-

Note

Poisons information/control centre: Hospital Carlos Finlay, Avenida 31 y Calle 114, La Habana [Head: Dr. R. Perez Cristia]

Responsible organisations

1. Registro Central de Plaguicidas
   Calle 150, No 2125
   e/21-A y 25,
   Siboney
   C.P. 12100
   LA HABANA

2. Ministerio de Salud Pública
   Calle N e/23 y 25, Vedado
   LA HABANA

3. Laboratorio de Toxicologia
   Insituto de Investigaciones de Sanidad Vegetal
   Calle 150 No. 2125 entre 21-A y 25, Siboney
   LA HABANA

4. Insituto de Medicina Veterinaria
   Calle 12, No. 355 e/15 y 17, Vedado
   LA HABANA

5. Instituto de Investigaciones para la Industria Alimenticia
   Km 6-1/2 Carretera del Guatao
   LA HABANA

6. Instituto Nacional de Investigaciones Fundamentales de Agricultura Tropical (INIFAT)
   Calle 1 esq 2,
   Santiago de las Vegas
   LA HABANA

7. Registro Central de Plaguicidas
   Calle 150,
   No. 2125 e/21-A y 25
   Rpto. Siboney, C.P. 12100
   LA HABANA
   Tel: (537) 219 665
   Fax: (537) 330 535, 335 086

8. Director de Relaciones Internatcionales
   Comité Estatal de Normalización
   Egido 610 e/Gloria y Apodaca
   C.P. 10100
   LA HABANA
   Tel: (53-7) 612068
   Fax: (53-7) 338048

9. El Director,
   Centro Nacional de Toxicologia (CENATOX)
   C.P. 14020
   Marianao 14
   LA HABANA

10. Dr. J.M.G. Fernandez
    CNC de Infoterra
    Director,
    Centro de Informacion, Divulgacion y Educacion Ambiental (CIDEA)
    Agencia de Medio Ambiente
    Ministerio de Ciencia, Technologia y Medio Ambiente
    Calle 18-A No. 4114 e/43 y 47
    Playa 11300
    LA HABANA
    Tel: (53 7) 296014, 201514
    Fax: (53 7) 330559

Cyprus
Kipriaki' Dimokratia, Kibris Cumhuriyeti, Republic of Cyprus

Responsibility	
Main authority responsible for approval/registration of pesticides	1, 5, 7
Main authority responsible for pesticide use directives, permits and training	1, 7
Main authority responsible for establishing national maximum pesticide residue limits	2
Main authority responsible for control of pesticide residues in food	3, 4, 7
Main laboratory for analysis of pesticide residues in food	3, 7
Main laboratory for quality control of pesticide products	7
Main organisation responsible for applied research regarding pesticide use. Information on related regional research	7
Designated national authority for notifications and information regarding banned and severely restricted pesticides including the PIC procedure	5
Codex Alimentarius contact point	6
National IRPTC correspondent	5, 7
Infoterra national focal point	8
National IFCS focal point	-
National association of pesticide manufacturers	9

Responsible organisations

1. Ministry of Agriculture,
Natural Resources and
Environment
NICOSIA
Tel: (2) 302 171, 303 225
Fax: (2) 445 156, 361 425

2. Ministry of Health
NICOSIA
Tel: (2) 303 243
Fax: (2) 303 498

3. State General Laboratory
Ministry of Health
NICOSIA
Tel: (2) 305 061
Fax: (2) 316 434

4. Ministry of Commerce,
Industry and Tourism
NICOSIA
Tel: (2) 303 441
Fax: (2) 366 120

5. The Chairman
Pest Control Products Board
Department of Agriculture
Ministry of Agriculture,
Natural Resources and
Environment
NICOSIA
Tel: (2) 302 254, 302 250
Fax: (2) 361 425

6. Dr I G Karis
Director,
Cyprus Organization for
Standards and Control of
Quality
Ministry of Commerce,
Industry and Tourism
NICOSIA

7. The Director
Department of Agriculture
Ministry of Agriculture,
Natural Resources and
Environment
L. Akritas Avenue
NICOSIA
Tel: (2) 302 250
Fax: (2) 361 425

8. Mr. N. Georgiades
Infoterra NFP
Environmental Services
Ministry of Agriculture,
Natural Resources and
Environment
L. Akritas Avenue
NICOSIA
Tel: (357 2) 302 883
Fax: (357 2) 363 945

9. Association of Importers-
Distributors of
Agrochemicals and
Fertilizers
P.O. Box 1455
NICOSIA
Tel: (2) 449 500
Fax: (2) 449 048

Czech Republic
Ceská Republika

Responsibility

Main authority responsible for approval/registration of pesticides	1, 2
Main authority responsible for pesticide use directives, permits and training	
Main authority responsible for establishing national maximum pesticide residue limits	3
Main authority responsible for control of pesticide residues in food	4, 5
Main laboratory for analysis of pesticide residues in food	2, 5
Main laboratory for quality control of pesticide products	
Main organisation responsible for applied research regarding pesticide use. Information on related regional research	3, 6, 7
Designated national authority for notifications and information regarding banned and severely restricted pesticides including the PIC procedure	8
Codex Alimentarius contact point	9
National IRPTC correspondent	10
Infoterra national focal point	12
National IFCS focal point	11
National association of pesticide manufacturers	-

Note

Poisons information/control centre: Clinic for Occupational Diseases, Vysehradska 49, Praha 2, Tel: 42 2 29 67 41, Fax: 42 2 26 65 03 [Head: Dr (Ms) D Pelclova]

Responsible organisations

1. Ministry of Agriculture
Tesnov 17
CZ-11705 PRAGUE 1

2. National Supervisory and Testing Institute of Agriculture
Pesticide and Pest Control Method Division
Zemedelská la
CZ-65837 BRNO
Fax: (42 5) 45211078

3. National Health Institute
National Reference Pesticide Center
Srobárova 48
CZ-10042 PRAGUE 10

4. National Veterinary Institute
Sídlistní 156 - Suchdol
CZ-16503 PRAGUE 6

5. National Food Quality Inspection
Jindricha Plachty 16
CZ-15119 PRAGUE 5

6. Crop Production Research Institute
Drnovská 507
CZ-16106 PRAGUE 6 - RUZYNE

7. Food Industry Research Institute
Radiová 7
CZ-10231 PRAGUE 10

8. Ministry of Environment
Vrsovická 65
10010 PRAGUE 10
Tel: (422) 67122759
Fax: (422) 67311611
(information exchange only)

9. Mr. P Dobrovsky
Codex Alimentarius Contact Point
Officer-in-charge
Odbor 320
Ministerstvo Zemedelstvi CR
Tesnov 17
CZ-11705 PRAGUE
Tel: (42 2) 2862869
Fax: (42 2) 2314117

10. Mr. V Mejstrik
Center of Ecology, Toxicology and Analysis
Research Institute for Organic Synthesis (CETA / VUOS)
543 18 PARDUBICE - RYBITVI
Tel: (4240) 42694
Fax: (4240) 43165

11. Mr Vladimir Bizek,
Deputy Minister of Environment
Vrsovicka 65
100 10 PRAGUE
Tel: (42 2) 6712 2836
Fax: (42 2) 6731 0308

12. Mrs. Milena Roudna
Infoterra NFP
International Relations Department
Ministry of the Environment
Vrsovicka 65
100 10 PRAGUE 10
Tel: (42 2) 6712 2769
Fax: (42 2) 6731 1611

Denmark
Danmark

Responsibility

Main authority responsible for approval/registration of pesticides	1, 2
Main authority responsible for pesticide use directives, permits and training	
Main authority responsible for establishing national maximum pesticide residue limits	3
Main authority responsible for control of pesticide residues in food	3
Main laboratory for analysis of pesticide residues in food	3
Main laboratory for quality control of pesticide products	
Main organisation responsible for applied research regarding pesticide use. Information on related regional research	4, 5, 7, 8
Designated national authority for notifications and information regarding banned and severely restricted pesticides including the PIC procedure	1
Codex Alimentarius contact point	9
National IRPTC correspondent	10
Infoterra national focal point	11
National IFCS focal point	12
National association of pesticide manufacturers	6

Note

Poisons information/control centre: Giftinformationscentralen, Rigshospitalet, Arbejdsmedicinsk Klinik 7122, Tangsvej 20, DK-2200 Copenhagen, Tel: +45-35-453545, Fax: +45-31-394233 [Head: Dr F Gyntelberg]

Resonsible organisations

1. Pesticides Division
   Danish Environmental
   Protection Agency
   Ministry of Environment and
   Energy
   Strandgade 29
   DK-1401 COPENHAGEN K
   Tel: (45 32) 660 100
   Fax: (45 32) 660 479

2. Kemikalieinspektionen
   Strandgade 29
   DK-1401 COPENHAGEN K

3. Levnedsmiddelstyrelsen
   Mørkhøj Bygade 19
   DK-2860 SØBORG
   Tel: (1) 696 600
   Fax: (1) 696 126

4. Statens Plantevaerncenter
   Flakkebjerg
   DK-4200 SLAGELSE
   Tel: (3) 535 863 00
   Fax: 535 863 71
   (Herbicides)

5. Statens Plantevaernscenter
   Lottenborgvej 2
   DK-2800 LYNGBY
   Tel: (2) 872 510
   (Fungicides and Insecticides)

6. Danish Agrochemical
   Association (DAA)
   Amalievej 20
   DK-1875
   FREDRIKSBORG C
   Tel: (45 31) 244 266
   Fax: (45 33) 258 416

7. Statens Skadedyrlaboratorium
   Skovbrynet 14
   DK-2800 LYNGBY
   Tel: (2) 878 055

8. National Agency for
   Environmental Protection
   Office of Chemicals and
   Pesticides
   Ministry of the Environment
   Strandgade 29
   DK-1401 COPENHAGEN K
   Tel: 315 78310
   Fax: 315 72449

9. Danish Codex Alimentarius
   Committee
   Danish Veterinary Service
   Rolighedsvej 25
   DK-1958
   FREDERIKSBERG C
   Tel: (45 31) 358 100
   Fax: (45 35) 361 912

10. The Director
    National Agency of
    Environmental Protection
    Attn: Library and
    Documentation Section
    Strandgade 29
    DK-1401 COPENHAGEN K

11. Ms. Dorte Bennedbaek
    Infoterra NFP
    Department of Environmental
    Data and Information
    Miljøstyrelsen
    Strandgade 29
    DK-1401 COPENHAGEN K
    Tel: (45 32) 660 100
    Fax: (45 32) 660 479

12. Ms L. Seedorff
Chemicals Division
Danish Environmental
Protection Agency
Strandgade 29
DK-1401 COPENHAGEN K
Tel: (45 32) 660100
Fax: (45 32) 660479

Djibouti

Jumhuriyah Jibuti, Republique de Djibouti

Responsibility	
Main authority responsible for approval/registration of pesticides	
Main authority responsible for pesticide use directives, permits and training	
Main authority responsible for establishing national maximum pesticide residue limits	
Main authority responsible for control of pesticide residues in food	
Main laboratory for analysis of pesticide residues in food	
Main laboratory for quality control of pesticide products	
Main organisation responsible for applied research regarding pesticide use. Information on related regional research	
Designated national authority for notifications and information regarding banned and severely restricted pesticides including the PIC procedure	-
Codex Alimentarius contact point	-
National IRPTC correspondent	-
Infoterra national focal point	1
National IFCS focal point	-
National association of pesticide manufacturers	-

Responsible organisations

1. M. Anis Abdallah
   PFN d'Infoterra
   Director,
   Institut Supérieur d'études et
   des recherches scientifiques
   et techniques
   B.P. 486
   DJIBOUTI
   Tel: (253) 352795
   Fax: (253) 354812

Dominica
Commonwealth of Dominica

Responsibility	
Main authority responsible for approval/registration of pesticides	
Main authotity responsible for pesticide use directives, permits and training	
Main authority responsible for establishing national maximum pesticide residue limits	
Main authority responsible control of pesticide residues in food	
Main laboratory for analysis of pesticide residues in food	
Main laboratory for quality control of pesticide products	
Main organisation responsible for applied research regarding pesticide use. Information on related regional research	
Designated national authority for notifications and information regarding banned and severely restricted pesticides including the PIC procedure	1
Codex Alimentarius contact point	-
National IRPTC correspondent	-
Infoterra national focal point	2
National IFCS focal point	3
National association of pesticide manufacturers	

Responsible organisations

1. The Director
Pesticide Control Board
Ministry of Agriculture
Botanic Gardens
ROSEAU
Tel: (809 44) 82401

2. Mr. Felix Gregoire
Infoterra NFP
Director,
Forestry and Wildlife Division
Ministry of Agriculture
Botanical Gardens
ROSEAU
Tel: (809) 4482401
Ext 415, 417
Fax: (809) 4487999

3. Permanent Secretary
Government Headquarters
Ministry of Agriculture and
Environment
ROSEAU
Fax: (809) 448 7999

Dominican Republic
República Dominicana

Responsibility

Main authority responsible for approval/registration of pesticides	-
Main authority responsible for pesticide use directives, permits and training	
Main authority responsible for establishing national maximum pesticide residue limits	-
Main authority responsible for control of pesticide residues in food	1
Main laboratory for analysis of pesticide residues in food	-
Main laboratory for quality control of pesticide products	
Main organisation responsible for applied research regarding pesticide use. Information on related regional research	-
Designated national authority for notifications and information regarding banned and severely restricted pesticides including the PIC procedure	2
Codex Alimentarius contact point	1
National IRPTC correspondent	3
Infoterra national focal point	-
National IFCS focal point	-
National association of pesticide manufacturers	4

Responsible organisations

1. Secretaria de Estado de Salud Pública y Asistencia Social
   Sección de Control de Alimentos
   Ensanche La Fe
   SANTO DOMINGO
   Tel: 565 3131

2. Director
   Ingeniero Agrónomo
   Depto. de Sanidad Vegetal
   Secretaria de Estado de Agricultura
   Los Jardinos
   SANTO DOMINGO
   Tel: 547 3786

3. El Director
   Comisión Ambiental
   Dr Delgado No 58-Apartado 1351
   SANTO DOMINGO

4. Asociación de Fabricantes e Importadores de Productos Agroquímicos (AFIPA)
   Apartado Postal 30158
   SANTO DOMINGO
   Tel: (809 56) 55344
   Fax: (809 56) 21984

Ecuador
República del Ecuador

Responsibility	
Main authority responsible for approval/registration of pesticides	1
Main authority responsible for pesticide use directives, permits and training	1
Main authority responsible for establishing national maximum pesticide residue limits	1, 2
Main authority responsible for control of pesticide residues in food	1
Main laboratory for analysis of pesticide residues in food	3
Main laboratory for quality control of pesticide products	3
Main organisation responsible for applied research regarding pesticide use. Information on related regional research	1, 4
Designated national authority for notifications and information regarding banned and severely restricted pesticides including the PIC procedure	1
Codex Alimentarius contact point	5
National IRPTC correspondent	1
Infoterra national focal point	6
National IFCS focal point	-
National association of pesticide manufacturers	7

Responsible organisations

1. Servicio Ecuatoriano de Sanidad Agropecuaria (SESA)
   Ministerio de Agricultura y Ganadería
   P.O. Box 1934 Suc.21
   QUITO
   Tel: (593 2) 567 232
   Fax: (593 2) 543 319

2. Ministerio de Salud Pública
   Calle Juan Larrea No. 444 y Riofrio
   QUITO
   Tel: (593 2) 528 745, 521 411
   Fax: (593 2) 569 786

3. Servicio Ecuatoriano de Sanidad Agropecuaria (SESA)
   Division Laboratorios
   (Lab. Plaguicidas)
   Ministerio de Agricultura y Ganaderia
   P.O. Box 1934, Suc.21
   QUITO
   Tel: (593 2) 567 232, 543 319
   Fax: (593 2) 228 448

4. Instituto Ecuatoriano de Investigaciones Agropecuarias (INIAP)
   Av. Eloy Alfaro y Amazonas
   P.O. Box 17171362
   QUITO
   Tel: (593 2) 567 645
   Fax: (593 2) 504 240

5. Instituto Ecuatoriano de Normalización (INEN)
   Calle Baquerizo Moreno No.454 y Av. 6 de Diciembre, Apartado 3999
   QUITO
   Tel: (593 2) 524 499, 565 626
   Fax: (593 2) 567 815

6. Dr. P L Paredes
   Secretario General de Planificacion
   Consejo Nacional de Desarollo (CONADE)
   Infoterra CNC
   Arenas y Manuel Larrea Edificio del Consejo Provincial de Pichincha
   QUITO
   Tel: (593 2) 520 017, 563 666
   Fax: (593 2) 563 002

7. Asociación de Importadores y Fabricantes de Insumos Agropecuarios (AIFA)
   Escobedo No. 631 y P. Solano, Piso 3, Oficina 303
   Casilla Postal 09018583
   GUAYAQUIL
   Tel: (593 4) 309 912
   Fax: (593 4) 562 641

Egypt

Jumhuriyah Misr Al-Arabiyah

Responsibility	
Main authority responsible for approval/registration of pesticides	1
Main authority responsible for pesticide use directives, permits and training	
Main authority responsible for establishing national maximum pesticide residue limits	1
Main authority responsible for control of pesticide residues in food	1, 2
Main laboratory for analysis of pesticide residues in food	1, 2, 7
Main laboratory for quality control of pesticide products	9
Main organisation responsible for applied research regarding pesticide use. Information on related regional research	1, 8
Designated national authority for notifications and information regarding banned and severely restricted pesticides including the PIC procedure	3
Codex Alimentarius contact point	4
National IRPTC correspondent	3
Infoterra national focal point	5
National IFCS focal point	6
National association of pesticide manufacturers	10

Note

Poisons information/control centres in Alexandria and Cairo

Responsible organisations

1. Ministry of Agriculture
Agricultural Research Centre
8 Gemeat El Kahira Street
GIZA
Tel: 721 128, 723 609,
723 000

2. General Laboratories
Administration
Ministry of Health
Magles El Shaab Street
Kasr El-Eini
CAIRO
Tel: 354 9696, 354 8544,
354 7371

3. The Chairman
Egyptian Environment Affairs
Agency(EEAA)
Cabinet of Ministers
17 Tiba Street
El-Mohandesen/Dokki
GIZA
Tel: (202) 360 1839, 360 1326
Fax: (202) 361 0764

4. The President
Egyptian Organization for
Standardization (EOS)
2 Latin America Street
Garden City
CAIRO
Tel: (21 2) 354 9720

5. Infoterra NFP
The Director General,
Executive Office for
Environmental Information
Academy for Scientific
Research and Technology
101 Kasr El Aini Street
CAIRO
Tel: (20 2) 354 0093
Fax: (20 2) 356 2820

6. Egyptian Environment Affairs
Agency
17 Teiba St.
El Mohandesseen
GIZA
Tel: (202) 360 119,360 1243,
360 1326
Fax: (202) 361 0764

7. Central Laboratory for
Pesticides
Ministry of Agriculture
Nadi El Sid Str.
Dokki
GIZA

8. Technical Cooperation Office
for the Environment
16 Lebnan Street
El Mohandesseen
GIZA
Tel: 302 9102, 302 6925, 303
1833

9. Chemical Pesticides
Department
National Research Center
Tahrir Street
Dokki
GIZA

10. Chemical Industry
Egyptian Holding Co.
5, Tolombat Garden City
CAIRO
Tel: (20 2) 235 4006
Fax: (20 2) 355 7475

El Salvador
República de El Salvador

Responsibility	
Main authority responsible for approval/registration of pesticides	1
Main authority responsible for pesticide use directives, permits and training	
Main authority responsible for establishing national maximum pesticide residue limits	1, 2
Main authority responsible for control of pesticide residues in food	3, 4
Main laboratory for analysis of pesticide residues in food	4
Main laboratory for quality control of pesticide products	
Main organisation responsible for applied research regarding pesticide use. Information on related regional research	4
Designated national authority for notifications and information regarding banned and severely restricted pesticides including the PIC procedure	5
Codex Alimentarius contact point	6
National IRPTC correspondent	7
Infoterra national focal point	8
National IFCS focal point	-
National association of pesticide manufacturers	9

Responsible organisations

1. Dirección General de Defensa Agropecuaria
P.O. Box 2265
SAN SALVADOR

2. Ministerio de Salud Pública y Asistencia Social
Calle Arce No 827
SAN SALVADOR

3. Laboratorio del Centro de Desarrollo Ganadero
P.O. Box 2265
SAN SALVADOR

4. Laboratorio de Centro Nacional de Tecnología Agricola (CENTA)
P.O. Box 2265
SAN SALVADOR

5. Jefe,
Division de Registro y Fiscalizacion
Direccion General de Sanidad Vegetal y Animal
Ministerio de Agricultura y Ganaderia
P.O. Box 554
SAN SALVADOR
Tel: (503) 294 0585
Fax: (503) 294 0585

6. Dr. R Molins
Gerente de Calidad Integral
Fundacion Salvadorena para el Desarollo Economico Social (FUSADES)
c/o CONACYT
9a Calle Poniente No. 4113
Colonia Escalon
SAN SALVADOR
Tel: (503) 237 036
Fax: (503) 237 036

7. Jefe de Laboratorios
Ministerio de Salud Pública y Asistencia Social
9a Ave Norte 120
SAN SALVADOR
Tel: 711507, 711523

8. Lic. Miguel Araujo
CNC de Infoterra
Secretario Ejecutivo del Medio Ambiente (SEMA)
Final 91 Avenida Norte, entre 11 y 13 Calle Poniente
Cos. Escalon
SAN SALVADOR
Tel: (503) 243 886, 793 830
Fax: (503) 239 083

9. Asociation de Proveedores Agricolas (APA)
Condominio Metro 2000, Local C-23
47 avenida Norte y Alameda Norte
Juan
Pablo II
SAN SALVADOR
Tel: (503) 791 017
Fax: (503) 791 017

Equatorial Guinea
Républica de Guinea Ecuatorial

Responsibility

Main authority responsible for approval/registration of pesticides	
Main authority responsible for pesticide use directives, permits and training	
Main authority responsible for establishing national maximum pesticide residue limits	
Main authority responsible for control of pesticide residues in food	
Main laboratory for analysis of pesticide residues in food	
Main laboratory for quality control of pesticide products	
Main organisation responsible for applied research regarding pesticide use. Information on related regional research	
Designated national authority for notifications and information regarding banned and severely restricted pesticides including the PIC procedure	-
Codex Alimentarius contact point	1
National IRPTC correspondent	-
Infoterra national focal point	-
National IFCS focal point	-
National association of pesticide manufacturers	

Responsible organisations

1. Mr. A. Ndjoli Mediko
   Jefe Nacional de Estadisticas
   Agropecuarias
   Ministerio de Agricultura,
   Ganaderia, Pesca y Forestal
   MALABO (Bioko Norte)

Eritrea

Responsibility

Main authority responsible for approval/registration of pesticides	
Main authority for pesticide use directives, permits and training	
Main authority responsible for establishing national maximum pesticide residue limits	
Main authority responsible for control of pesticide residues in food	
Main laboratory for analysis of pesticide residues in food	
Main laboratory for quality control of pesticide products	
Main organisation responsible for applied research regarding pesticide use. Information on related regional research	
Designated national authority for notifications and information regarding banned and severely restricted pesticides including the PIC procedure	-
Codex Alimentarius contact point	-
National IRPTC correspondent	-
Infoterra national focal point	1
National IFCS focal point	2
National association of pesticide manufacturers	-

Responsible organisations

1. Infoterra N F P
   Ministry of Agriculture
   P.O. Box 1048
   ASMARA
   Tel: (291 1) 181077
   Fax: (291 1) 181415

2. Mr. N. Gebremedhin
   Acting Executive Director
   Eritrean Agency forthe
   Environment
   P.O. Box 5713
   ASMARA
   Tel: (2911) 181 324
   Fax: (2911) 182 418

Estonia

Eesti Vabariik

Responsibility

Main authority responsible for approval/registration of pesticides	1
Main authority responsible for pesticide use directives, permits and training	5
Main authority responsible for establishing national maximum pesticide residue limits	7
Main authority responsible for control of pesticide residues in food	7
Main laboratory for analysis of pesticide residues in food	6, 7
Main laboratory for quality control of pesticide products	6
Main organisation responsible for applied research regarding pesticide use. Information on related regional research	8
Designated national authority for notifications and information regarding banned and severely restricted pesticides including the PIC procedure	1
Codex Alimentarius contact point	2
National IRPTC correspondent	-
Infoterra national focal point	3
National IFCS focal point	4
National association of pesticide manufacturers	-

Responsible organisations

1. State Department of Plant Protection
Division of Pesticides Registration
Teaduse 2
EE-3400 SAKU-HARJUMAA
Tel: (372 2) 721 895
Fax: (372 2) 721 662

2. National Board of Food
Attn: Ms. Kaja Kuivjogi
Ministry of Social Affairs
Gonsiori 29
EE-0001 TALLINN
Tel: (372 2) 421 267
Fax: (372 2) 426 268

3. Infoterra NFP/Estonia
Attn: Mr. Leo Saare
Estonian Environmental Information Centre
Ministry of Environment
Mustamäe tee 33
EE-0006 TALLINN
Tel: (372 2) 639 4151
Fax: (372 2) 639 4071

4. National Working Environment Board of Estonia
5, Estonia Avenue
EE-0001 TALLINN
Tel:(372 2) 440 854
Fax:(372 2) 455 111

5. State Department of Plant Protection
Teaduse 2
EE-3400 SAKU-HARJUMAA
Tel: (372 2) 721 700
Fax: (372 2) 721 662

6. Division of Environmental Research
State Department of Plant Protection
Teaduse 4/6
EE-3400 SAKU-HARJUMAA
Tel: (372 2) 721 705,
721 784
Fax: (372 2) 721 074

7. National Board of Health Protection
Paldiski Road 81
EE-0006 TALLINN
Tel: (372 2) 491 752
Fax: (372 2) 476 051

8. Department of Plant Protection
Estonian Research Institute of Agriculture
Teaduse 4/6
EE-3400 SAKU-HARJUMAA
Tel: (372 2) 721 879
Fax: (372 2) 771 385

Ethiopia
Ityop'iya

Responsibility	
Main authority responsible for approval/registration of pesticides	1, 2, 3
Main authority responsible for pesticide use directives, permits and training	
Main authority responsible for establishing national maximum pesticide residue limits	1, 3, 4
Main authority responsible for control of pesticide residues in food	4, 5, 6
Main laboratory for analysis of pesticide residues in food	5, 6, 7, 8
Main laboratory for quality control of pesticide products	
Main organisation responsible for applied research regarding pesticide use. Information on related regional research	3, 9, 10
Designated national authority for notifications and information regarding banned and severely restricted pesticides including the PIC procedure	12
Codex Alimentarius contact point	4
National IRPTC correspondent	-
Infoterra national focal point	11
National IFCS focal point	-
National association of pesticide manufacturers	-

Responsible organisations

1. Ministry of Agriculture
P.O Box 2347
ADDIS ABABA
Tel: 448 040

2. Ministry of Health
P.O. Box 1234
ADDIS ABABA
Tel: 157 011, 157 102, 157 831

3. Ministry of State Farms Development
P.O Box 5765
ADDIS ABABA
Tel: 154 660

4. Ethiopian Standards Institution (ESI)
P.O. Box 2310
ADDIS ABABA
Tel: 150 400, 150 425

5. Ethiopian Nutrition Institution (ENI)
P.O. Box 5654
ADDIS ABABA
Tel: 151 600

6. National Research Institution of Health
P.O. Box 1242
ADDIS ABABA
Tel: 133 470

7. Department of Chemistry
Addis Ababa University
P.O. Box 1176
ADDIS ABABA
Tel: 113 177

8. Sholla Plant Protection Laboratory
Pesticide Chemistry Unit
ADDIS ABABA

9. Phyto-Pathology Laboratory
Ambo Phyto-Biology Research Centre
P.O. Box 37
AMBO

10. Institute of Agricultural Research
P.O. Box 2003
ADDIS ABABA
Tel: 161 055

11. Infoterra NFP
Documentation and Information
National Environmental Protection Authority
P.O. Box 12 760
ADDIS ABABA
Tel: (251 1) 152 468, 182 967
Fax: (251 1) 513 042

12. Plant Production and Protection Department
Ministry of Agriculture
P.O. Box 2347
ADDIS ABABA
Fax: (251 1) 611 648

Fiji
Sovereign Democratic Republic of Fiji

Responsibility	
Main authority responsible for approval/registration of pesticides	1
Main authority responsible for pesticide use directives, permits and training	1, 7
Main authority responsible for establishing national maximum pesticide residue limits	-
Main authority responsible for control of pesticide residues in food	2, 3
Main laboratory for analysis of pesticide residues in food	-
Main laboratory for quality control of pesticide products	-
Main organisation responsible for applied research regarding pesticide use. Information on related regional research	7
Designated national authority for notifications and information regarding banned and severely restricted pesticides including the PIC procedure	4
Codex Alimentarius contact point	5
National IRPTC correspondent	1
Infoterra national focal point	6
National IFCS focal point	1
National association of pesticide manufacturers	-

Responsible organisations

1. Registrar of Pesticides
   Koronivia Research Station
   P.O. Box 77
   NAUSORI
   Tel: (679) 477 044
   Fax: (679) 400 262

2. Chief Health Inspector
   Central Board of Health
   Ministry of Health and Social Welfare
   Government Buildings
   SUVA
   Tel: (679) 306 177
   Fax: (679) 306 163

3. Principal Research Officer (Chemistry)
   Koronivia Research Station
   P.O. Box 77
   NAUSORI
   Tel: (679) 477 044
   Fax: (679) 400 262

4. The Deputy Permanent Secretary Services
   Ministry of Agriculture, Fisheries and Forests
   G.P.O. Box 358
   SUVA
   Tel: (679) 311 233
   Fax: (679) 302 478

5. The Permanent Secretary
   Ministry of Agriculture, Fisheries, Forests and ALTA
   P.O. Box 358
   SUVA
   Tel: (679) 311 233
   Fax: (679) 302 478

6. Infoterra NFP
   Town and Country Planning Office
   Governmental Buildings
   P.O. Box 2350
   SUVA
   Tel: (679) 211 717
   Fax: (679) 303 515

7. Research Division
   Ministry of Agriculture, Fisheries, Forests and ALTA
   P.O. Box 77
   NAUSORI
   Tel: (679) 477 044
   Fax: (679) 400 262

Finland

Suomen Tasavalta

Responsibility	
Main authority responsible for approval/registration of pesticides	1, 2, 3, 4
Main authority responsible for pesticide use directives, permits and training	
Main authority responsible for establishing national maximum pesticide residue limits	3, 5
Main authority responsible for control of pesticide residues in food	6, 7, 8 9, 10, 11
Main laboratory for analysis of pesticide residues in food	7, 8, 9, 10
Main laboratory for quality control of pesticide products	
Main organisation responsible for applied research regarding pesticide use. Information on related regional research	12, 13
Designated national authority for notifications and information regarding banned and severely restricted pesticides including the PIC procedure	4
Codex Alimentarius contact point	14
National IRPTC correspondent	17
Infoterra national focal point	15
National IFCS focal point	18
National association of pesticide manufacturers	16

Note

Poisons information/control centre: Children's Hospital, Stenbäckinkatu 11, SF-00290 HELSINKI, Tel: +358-47-112788, Fax: +358-47-14702 [Head: Dr J Vilska]

Responsible organisations

1. Pesticide Commission
   c/o Plant Production Inspection Centre
   Plant Protection Department
   P.O. Box 42
   FIN-00501 HELSINKI
   Tel: (358 0) 134 211
   Fax: (358 0) 134 214 21

2. Plant Production Inspection Centre
   Plant Protection Department
   P.O. Box 42
   FIN-00501 HELSINKI
   Tel: (358 0) 134 211
   Fax: (358 0) 134 214 99

3. Ministry of Social Affairs and Health
   P.O. Box 197
   FIN-00531 HELSINKI
   Tel: (358 0) 160 5340
   Fax: (358 0) 160 5343

4. Finnish Environment Institute
   Chemicals Division
   P.O. Box 140
   FIN-00251 HELSINKI
   Tel: (358 0) 403 000
   Fax: (358 0) 403 005 91

5. Ministry of Trade and Industry
   Aleksanterinkatu 4
   FIN-00170 HELSINKI
   Tel: (358 (0) 160 3553
   Fax: (358 0) 160 3666

6. National Food Administration
   P.O. Box 5
   FIN-00531 HELSINKI
   Tel: (358 0) 772 676 16
   Fax: (358 0) 772 676 66

7. Customs Laboratory
   Tekniikantie 13
   FIN-02150 ESPOO
   Tel: (358 0) 614 3267
   Fax: (358 0) 463 383

8. Plant Production Inspection Centre
   Agricultural Chemistry Department
   P.O. Box 83
   FIN-01301 VANTAA
   Tel: (358 0) 857 056 00
   Fax: (358 0) 831 198

9. Technical Research Centre
   Food Research Laboratory
   Biologinkuja 1
   FIN-02150 ESPOO
   Tel: (358 0) 4561
   Fax: (358 0) 455 2103

10. National Veterinarian and Food Research Institute
    Hämeentie 57
    FIN-00550 HELSINKI
    Tel: (358 0) 393 101
    Fax: (358 0) 393 1811

11. Helsinki City Environmental Laboratory
    Helsinginkatu 24
    FIN-00530 HELSINKI
    Tel: (358 0) 70991
    Fax: 358(0) 709 922 45

12. Forest Research Institute
    P.O. Box 18
    FIN-01301 VANTAA
    Tel: (358 0) 857 051
    Fax: (358 0) 857 2575

13. Agricultural Research Centre
Institute of Plant Protection
FIN-31600 JOKIOINEN
Tel: (358 16) 1881
Fax: (358 16) 188 584

14. Ministry of Trade and Industry
Advisory Committee on Foodstuffs
Secretary General
P.O. Box 230
FIN-00170 HELSINKI
Tel: (358 (0) 160 3553
Fax: (358 0) 160 3666

15. Infoterra NFP
Ministry of the Environment
Environmental Protection Department
P.O. Box 399
FIN-00121 HELSINKI
Tel: (358 0) 19911
Fax: (358 0) 199 1499

16. Agrochemical Producers' Association of Finland
P.O. Box 4
FIN-00131 HELSINKI
Tel: (358 0) 172 841
Fax: (358 0) 579 052, 630 225

17. Ministry of the Environment
Environmental Protection Department
Attn: Ms Ann-Britt Ylinen
P.O. Box 399
FIN-00121 HELSINKI
Tel: (3580) 19911

18 Ms Pirkko Kivelä-Ikonen
Ministry of the Environment
Environmental Protection Department
P.O. Box 399
FIN-00121 HELSINKI
Tel: (358 0) 1991 9733
Fax: (358 0) 1991 9716

France

République Française

Responsibility	
Main authority responsible for approval/registration of pesticides	1, 2
Main authority responsible for pesticide use directives, permits and training	
Main authority responsible for establishing national maximum pesticide residue limits	1, 3, 4, 5 6
Main authority responsible for control of pesticide residues in food	7, 8, 9
Main laboratory for analysis of pesticide residues in food	7, 8, 9, 10
Main laboratory for quality control of pesticide products	
Main organisation responsible for applied research regarding pesticide use. Information on related regional research	12
Designated national authority for notifications and information regarding banned and severely restricted pesticides including the PIC procedure	13, 14
Codex Alimentarius contact point	15
National IRPTC correspondent	16
Infoterra national focal point	17
National IFCS focal point	18
National association of pesticide manufacturers	11

Note

Poisons information/control centres in Angers, Bordeaux, Clermont-Ferrand, Grenoble, Lille, Lyon, Marseille, Montpellier, Nancy, Paris, Reims, Rennes, Rouen, Strasbourg, Toulouse and Tours

Responsible organisations

1. Commission d'Etude de la Toxicité des Produits Antiparasitaires à Usage Agricole et des Produits Assimilés - Secrétariat administratif
   175, rue du Chevaleret
   F-75646 PARIS CEDEX 13
   Tel: 49558139
   Fax: 49555949

2. Ministère de l'Agriculture et de la Pêche
   Direction Générale de l'Alimentation
   Sous Direction de la Protection des Végétaux
   F-75646 PARIS CEDEX 13
   Tel: 49558157
   Fax: 49555949

3. Ministère des Affaires Sociales, de la Santé et de la Ville
   1 Place Fontenoy
   F-75350 PARIS 07
   Tel: 40566000

4. Direction Générale de l'Alimentation
   175 rue de Chevaleret
   F-75646 PARIS CEDEX 13
   Tel: 49555810

5. Direction de la Concurrence, de la Consommation et de la Répression de Fraudes
   59 Bd Vincent Auriol
   F-75703 PARIS CEDEX 13
   Tel: 44871717

6. Conseil Supérieur d'Hygiène Publique de France
   1, Place Fontenoy
   75350 PARIS 07
   Tel: 40566000

7. C.N.E.V.A.
   Laboratoire Central d'Hygiène Alimentaire
   43 rue de Dantzig
   F-75015 PAIS
   Tel: 45311480
   Fax: 45312994

8. D.G.C.C.R.F.
   Laboratoire Central de Recherches et d'Analyses
   25 Avenue de la République
   F-91035 MASSY
   Tel: 69206280
   Fax: 69206763

9. Institut National de Recherche Agricole (INRA)
   Laboratoire de Phytopharmacie
   Route de Saint Cyr
   F-78026 VERSAILLES
   Tel: 30833000
   Fax: 30833119

10. Faculté de Pharmacie
    Laboratoire de Chimie Appliquée à l'Expertise
    15, Avenue Charles Flahault
    F-340 60 MONTPELLIER CEDEX
    Tel: 67635360

11. Union des Industries de la Protection des Plantes (UIPP)
2 rue Denfert Rocherau
F-92100 BOULOGNE-BILLANCOURT
Tel: (33 1) 4605 5052
Fax: (33 1) 4825 1688

12. Association de Coordination Technique Agricole (ACTA)
149 rue de Bercy
F-75595 PARIS CEDEX 12
Tel: 40045000
Fax: 40045011

13. Monsieur le Directeur,
Sous-Direction de la Protection des Végétaux
Bureau de Controle des Produits Phytosanitaires
Ministère de l'Agriculture et de la Fôret
175, rue du Chevaleret
F-75646 PARIS CEDEX 13
Tel: (33 1) 158 413 13
Fax: (33 1) 457 081 69

14. Monsieur le Directeur
Direction de la prévention des pollutions
Mission du controle des produits
Ministère de l'environnement
20 Av. de Segur
F-75302 PARIS 07-S.P.
Tel: (33 1) 421 915 44
Fax: (33 1) 421 914 68

15. Ministère de l'Economie
D.G.C.C.R.F
Secrétariat General du Comite National du Codex Alimentarius
Attention: Mme. Michelle Gunzle
59 Bd Vincent Auriol
Teledoc 051
F-75703 PARIS CEDEX 13
Tel: 44972963
Fax: 44973043

16. Chef de la Mission du Contrôle des Produits
Direction de l'Eau, de la Prévention des Pollutions et des Risques
Attn: M.P. Deschamps
Ministère de l'Environnement
14 Bd du Général Leclerc
F-92524 NEUILLY SUR SEINE CEDEX
Tel: 40812122
Fax: 40811818

17. M. B. Morel
PFN d'Infoterra
Directeur,
IFEN
17 rue des Huguenots
F-45058 ORLEANS CEDEX
Tel: (33 38) 797878
Fax: (33 38) 797870

18. Direction generale de la Sante'
(Address unstated)

Gabon
République Gabonaise

Responsibility	
Main authority responsible for approval/registration of pesticides	1, 2, 3, 4
Main authority responsible for pesticide use directives, permits and training	
Main authority responsible for establishing national maximum pesticide residue limits	1, 4
Main authority responsible for control of pesticide residues in food	1, 2, 3, 5
Main laboratory for analysis of pesticide residues in food	-
Main laboratory for quality control of pesticide products	
Main organisation responsible for applied research regarding pesticide use. Information on related regional research	2, 3
Designated national authority for notifications and information regarding banned and severely restricted pesticides including the PIC procedure	-
Codex Alimentarius contact point	6
National IRPTC correspondent	7
Infoterra national focal point	8
National IFCS focal point	9
National association of pesticide manufacturers	-

Responsible organisations

1. Centre National Anti-Pollution (CENAP)
   B.P. 3241
   LIBREVILLE

2. Centre Universitaire des Science de la Santé (CUSS)
   LIBREVILLE

3. Centre International d'Amélioration des Cultures Maraîchères (CIAM)
   LIBREVILLE

4. Institut des Recherches Agronomiques et Forestières (IRAF)
   LIBREVILLE

5. Ministère de la Santé Publique et de la Population
   LIBREVILLE

6. Commission Nationale Gabonaise de la FAO
   Ministère de l'Agriculture, de l'Élevage et du Developpement Rurale
   B.P. 551
   LIBREVILLE
   Tel: 722561

7. Monsieur le Directeur
   Centre national anti-pollution
   Ministère de l'environnement et de la protection de la nature
   B.P. 3241
   LIBREVILLE

8. M.J. Marcel Malolas
   Directeur des études, du contentieux et du droit de l'environnement
   Point Focal PNUE/Infoterra
   Direction Générale de l'environnement et de la protection de la nature
   B.P. 3903
   LIBREVILLE
   Tel: (241) 731707, 730873

9. Dr. D Mavoungou-Nolet
   Environmental Information Agent
   Concept International
   B.P. 3818
   LIBREVILLE
   Tel: (241) 74 87 36/72 78 63
   Fax: (241) 74 87 37

Gambia

Republic of The Gambia

Responsibility	
Main authority responsible for approval/registration of pesticides	1, 7
Main authority responsible for pesticide use directives, permits and training	
Main authority responsible for establishing national maximum pesticide residue limits	1
Main authority responsible for control of pesticide residues in food	1
Main laboratory for analysis of pesticide residues in food	2, 3
Main laboratory for quality control of pesticide products	
Main organisation responsible for applied research regarding pesticide use. Information on related regional research	4
Designated national authority for notifications and information regarding banned and severely restricted pesticides including the PIC procedure	7
Codex Alimentarius contact point	6
National IRPTC correspondent	9
Infoterra national focal point	5
National IFCS focal point	8
National association of pesticide manufacturers	-

Responsible organisations

1. Hazardous Chemicals and Pesticides Board
   National Environment Agency
   5 Marina Parade
   BANJUL
   Tel: (220) 28056
   Fax: (220) 29701

2. Pesticides Analytical Laboratory
   National Environment Agency
   5 Marina Parade
   BANJUL
   Tel: (220) 28056
   Fax: (220) 29701

3. Pesticide Residue Laboratory
   Department of Water Resources
   7 Marina Parade
   BANJUL
   Tel: (220) 82162
   Fax: (220) 28628

4. Agricultural Pest Management Unit
   YUNDUM
   Tel: (220) 82207
   Fax: (220) 95413

5. Ms. Ndey-Isatou Njie
   Infoterra NFP
   Executive Director,
   National Environment Agency
   P.M.B. 48
   BANJUL
   Tel: (220) 228 056
   Fax: (220) 229 701

6. The Director
   Department of Agriculture
   Central Bank Building
   Buckle Street
   BANJUL
   Tel: (220) 95420
   Fax: (220) 95413

7. Registrar of Pesticides and Chemicals
   National Environment Agency
   5 Fitzgerald Street
   PMB 18
   BANJUL
   Tel: (220) 228 056
   Fax: (220) 229 701

8. Mr Ndey-Isatou Njie
   Executive Director,
   National Environment Agency
   5 Fitzgerald Street, PMB 48
   BANJUL
   Tel: (220) 228 056
   Fax: (220) 229 701

9. The Director,
   Environment Unit
   Ministry of Health, the Environment and Social Welfare
   The Quadrangle
   BANJUL

Georgia
Sakartvelos Respublikis, Republic of Georgia, Gruzia

Responsibility

Main authority responsible for approval/registration of pesticides	
Main authority responsible for pesticide use directives, permits and training	
Main authority responsible for establishing national maximum pesticide residue limits	
Main authority responsible for control of pesticide residues in food	
Main laboratory for analysis of pesticide residues in food	
Main laboratory for quality control of pesticide products	
Main organisation responsible for applied research regarding pesticide use. Information on related regional research	
Designated national authority for notifications and information regarding banned and severely restricted pesticides including the PIC procedure	-
Codex Alimentarius contact point	-
National IRPTC correspondent	-
Infoterra national focal point	1
National IFCS focal point	-
National association of pesticide manufacturers	-

Responsible organisations

1. Mr. Grigori Abramia
Infoterra NFP
Head,
International Relations
Department
Ministry of Environmental
Protection
Kostava Str. 68a
TBILISI 380015
Tel: (7 8832) 983425

Germany
Bundesrepublik Deutschland

Responsibility

Main authority responsible for approval/registration of pesticides	1, 2, 3, 4, 5, 16
Main authority responsible for pesticide use directives, permits and training	
Main authority responsible for establishing national maximum pesticide residue limits	1, 2, 4, 5, 6
Main authority responsible for control of pesticide residues in food	6
Main laboratory for analysis of pesticide residues in food	3, 4, 6, 7
Main laboratory for quality control of pesticide products	
Main organisation responsible for applied research regarding pesticide use. Information on related regional research	3, 5
Designated national authority for notifications and information regarding banned and severely restricted pesticides including the PIC procedure	8, 9
Codex Alimentarius contact point	10
National IRPTC correspondent	11, 12
Infoterra national focal point	13
National IFCS focal point	15
National association of pesticide manufacturers	14

Note

Poisons information/control centres in Berlin, Bonn, Braunschweig, Bremen, Erfurt, Freiburg/Breisgau, Göttingen, Hamburg, Homburg/Saar, Kiel, Koblenz, Ludwigshafen, Mainz, München, Münster, Nürnburg and Papenburg

Responsible organisations

1. Bundesministerium für Ernährung, Landwirtschaft und Forsten
Rochusstrasse 1
D-53123 BONN
Tel: (228) 5290
Fax: (228) 529 4262

2. Bundesministerium für Gesundheit
Am Propsthof 78 a
D-53121 BONN
Tel: (228) 9410
Fax: (228) 941 4900

3. Biologische Bundesanstalt für Land- und Forstwirtschaft
Abteilung für Pflanzenschutzmittel und Anwendungstechnik
Messeweg 11/12
D-38104 BRAUNSCHWEIG
Tel: (49 531) 299 3400
Fax: (49 531) 299 3002

4. Bundesinstitut für Gesundheitlichen Verbraucherschutz und Veterinärmedizin (BgVV)
Thielallee 88-92
Postfach 330013
D-14191 BERLIN
Tel: (30) 84120
Fax: (30) 8412 3374

5. Fachbereich Pflanzenschutz- und Schädlings-bekämpfungsmittel (ZEBS) des BgVV
Thielallee 88-92
Postfach 330013
D-14191 BERLIN
Tel: (30) 8412 3291
Fax: (30) 8412 3894

6 Zentrale Erfassungs- und Bewertungsstelle für Umweltchemikalien (ZEBS) des BgVV
Thielallee 88-92
Postfach 330013
D-14191 BERLIN
Tel: (30) 8412 3340
Fax: (30) 8412 3359

7. Deutsche Forschungs-gemeinschaft
Senatskommission zur Beurteilung von Stoffen in der Landwirdschaft
Kennedyalle 40
D-53175 BONN
Tel: (228) 885 2341
Fax: (228) 885 2728

8. Bundesanstalt für Arbeitsschutz Anmeldestelle Chemikaliengesetz
Friedrich-Henkel-Weg 1-25
D-44149 DORTMUND
Tel: (49 231) 907 1515
Fax: (49 231) 907 1679

9. The Director
Biologische Bundesanstalt für Land- und Forstwirtschaft
Abteilung für Pflanzenschutzmittel und Anwendungstechnik
Koordinierungsgruppe
Messeweg 11/12
D-38104 BRAUNSCHWEIG
Tel: (49 531) 299 3452
Fax: (49 531) 299 3003

10. Ms Jutta Peters
Ministerialdirektorin
Bundesministerium für Gesundheit
Am Propsthof 78 a
D-53121 BONN
Tel: (0228) 941 4100
Fax: (0228) 941 4940

11. Umweltbundesamt
Bewertungsstelle Chemikaliengesetz
Bismarckplatz 1
D-14193 BERLIN
Tel: (49 30) 23145-5
Fax: (49 30) 23145-856

12. Fachbereich Chemikalienbewertung des BgVV
Thielallee 88-92
Postfach 330013
D-14191 BERLIN
Tel: (30) 8412 3781
Fax: (30) 8412 3851

13. Dr. K. Lüdcke
Infoterra NFP
Umweltbundesamt
Zentrale Fachbibliothek Umwelt
Bismarckplatz 1
D-14193 BERLIN
Tel: (49 30) 8903 2213
Fax: (49 30) 8903 2285

14. Industrieverband Agrar e.V. (IVA)
Karlstrasse 21
D-60329 FRANKFURT/MAIN Tel: (49 69) 255 6281
Fax: (49 69) 236 702

15. Dr. U Schlottman
Federal Ministry for Environment, Nature Conservation and Nuclear Safety
Postfach 120629
Kennedyallee 2
D-53175 BONN
Tel: (49 228) 305 2740
Fax: (49 228) 305 3524

16. Umweltbundesamt
Einvernehmensstelle Pflanzenschutzgesetz
Bismarckplatz 1
D-14193 BERLIN
Tel: (49 30) 23145
Fax: (49 30) 23145

Ghana
Republic of Ghana

Responsibility	
Main authority responsible for approval/registration of pesticides	1, 2
Main authority responsible for pesticide use directives, permits and training	
Main authority responsible for establishing national maximum pesticide residue limits	1
Main authority responsible for control of pesticide residues in food	1, 3, 4, 5, 6
Main laboratory for analysis of pesticide residues in food	1, 5
Main laboratory for quality control of pesticide products	
Main organisation responsible for applied research regarding pesticide use. Information on related regional research	7
Designated national authority for notifications and information regarding banned and severely restricted pesticides including the PIC procedure	8
Codex Alimentarius contact point	9
National IRPTC correspondent	8
Infoterra national focal point	8
National IFCS focal point	11
National association of pesticide manufacturers	10

Responsible organisations

1. Ghana National Standards Board
   P.O. Box M 245
   ACCRA
   Tel: (233 21) 776 231,
   776 191
   Fax: (233 21) 776 092

2. Environmental Protection Council (EPC)
   P.O. Box M 326
   ACCRA

3. Veterinary Services Division
   Ministry of Agriculture
   P.O. Box M 161
   ACCRA
   Tel: 76015

4. Ministry of Industries, Science and Technology
   P.O. Box M 39
   ACCRA
   Tel: 66875

5. Food Research Institute C.S.I.R.
   P.O. Box M 20
   ACCRA
   Tel: 77330,77647

6. Ministry of Agriculture
   P.O. Box M 37
   ACCRA
   Tel: 65421

7. Cocoa Research Institute
   P.O. Box 8
   TAFO, Eastern Region

8. The Executive Director,
   Environmental Protection Council
   P.O. Box M 326
   ACCRA
   Tel: (233 21) 664 697,
   662 626
   Fax: (233 21) 662 690

9. The Director
   Ghana Standards Board
   P.O. Box M 245
   ACCRA

10. Agrochemicals Association of Ghana
    c/o Tema Chemicals Ltd.
    P.O. Box 104
    TEMA
    Tel: (233 221) 2991
    Fax: (233 221) 712 946

11. Mr T Bruce
    Chief Inspector of Factories
    Department of Factories Inspectorate
    Ministry of Employment and Social Welfare
    P.O. Box M 58
    ACCRA
    Tel: (233 21) 662 633
    Fax: (233 21) 667 251

Greece
Ellinikí Dhimokratiá

Responsibility	
Main authority responsible for approval/registration of pesticides	1, 2
Main authority responsible for pesticide use directives, permits and training	2
Main authority responsible for establishing national maximum pesticide residue limits	1, 2, 3
Main authority responsible for control of pesticide residues in food	1, 2, 11, 12, 13, 14
Main laboratory for analysis of pesticide residues in food	1, 4
Main laboratory for quality control of pesticide products	1
Main organisation responsible for applied research regarding pesticide use. Information on related regional research	1, 5, 15
Designated national authority for notifications and information regarding banned and severely restricted pesticides including the PIC procedure	2
Codex Alimentarius contact point	7
National IRPTC correspondent	8
Infoterra national focal point	9
National IFCS focal point	10
National association of pesticide manufacturers	6

Note

Poisons information/control centres in Athens ant Thessaloniki

Responsible organisations

1. Benaki Phytopathological Institute
   8 Delta Street
   Kifissia
   GR-1456 ATHENS

2. Department of Pesticides
   Ministry of Agriculture
   3-5 Hippokratus Street
   GR-10679 ATHENS
   Tel: (30 1) 529 1412
   Fax: (30 1) 529 1413

3. Ministry of Agriculture
   Veterinary Service
   16 Kapnokoptiriou Street
   GR-10433 ATHENS

4. General Chemical State Laboratory
   16 an. Tsoha Street
   Ampelokipi
   GR-11521 ATHENS
   Tel: 64 66 917
   Fax: 64 65 123

5. Ministry of Agriculture
   Division of Agricultural Research
   15 Halkokondili Street
   GR-10432 ATHENS

6. Greek Agrochemical Association (GAA)
   53 Patission Avenue
   GR-10433 ATHENS
   Tel: (30 1) 522 1542
   Fax: (30 1) 522 1542

7. Ministry of Agriculture
   Direction of Processing Standardisation and Quality Control of Agricultural Products of Vegetable Origin
   2 Acharnon Street
   GR-10432 ATHENS
   Tel: 524 6364
   Fax: 524 0955

8. The Director
   University of Athens
   School of Medicine
   Department of Forensic Medicine and Toxicology
   18 Ipsilandou Colonaki
   GR-ATHENS

9. Mr Paris Zikos
   Ministry of the Environment
   Physical Planning and Public Works
   Environmental Division
   14728 Octovriou (Patission) Street
   GR-11251 ATHENS
   Tel: (30 1) 865 0053,
   865 0334

10. Mrs A Kafetzopoulos
    General Director,
    Directorate of Working Conditions and Health
    Ministry of Labour
    40 Pireos Street
    10182 ATHENS
    Tel: 523 0377
    Fax: 524 3385

11. Regional Centre for Plant Protection and Quality Control of Piraeus
Laboratory for the Control of Pesticide Residues
Attn: Dr.(Mrs) L. Rizou
1 Sofokli Venizelou Str.
14123 LICOVRISI, ATTIKIS
Tel: (30 1) 284 0740
Fax: (30 1) 281 9732

12. Regional Centre for Plant Protection and Quality Control of Patra
Laboratory for the Control of Pesticide Residues
Attn: Mr C Markopoulos
4 Gerokostopoulou Street
26221 PATRA
Tel: (30 61) 270 912
Fax: (30 61) 623 238

13. Regional Centre for Plant Protection and Quality Control of Thessaloniki
Laboratory for the Control of Pesticide Residues
Attn: Dr. N. Niklis
1 Chapsa Street
P.O. Box 14516
54626 THESSALONIKI
Tel: (30 31) 471 210
Fax: (30 31) 471 210

14. Regional Centre for Plant Protection and Quality Control of Iraklio
Laboratory for the Control of Pesticide Residues
Attn: Mrs Maria Mentzaki
17 Monis Grevernetou Str.
71307 IRAKLIO, KRITI
Tel: (30 81) 229 448
Fax: (30 81) 225 616

15. National Agricultural Research Foundation
Egialias 19 and Chalepa
15125 MAROUSSI
Tel: (30 1) 684 0333
Fax: (30 1) 684 6700

Grenada

Responsibility	
Main authority responsible for approval/registration of pesticides	2, 3
Main authority responsible for pesticide use directives, permits and training	
Main authority responsible for establishing national maximum pesticide residue limits	1, 2, 4
Main authority responsible for control of pesticide residues in food	2, 4, 5
Main laboratory for analysis of pesticide residues in food	1,2
Main laboratory for quality control of pesticide products	
Main organisation responsible for applied research regarding pesticide use. Information on related regional research	2, 4, 5
Designated national authority for notifications and information regarding banned and severely restricted pesticides including the PIC procedure	2
Codex Alimentarius contact point	1
National IRPTC correspondent	-
Infoterra national focal point	-
National IFCS focal point	6
National association of pesticide manufacturers	-

Responsible organisations

1. Dr. Peter Radix
   Director,
   Grenada Bureau of Standards
   c/o Produce Chemist
   Laboratory
   Tyrell Street
   ST. GEORGE'S
   Tel: 440 5886,
   440 6783
   Fax: 440 4115

2. Grenada Pesticide Control
   Board
   c/o Produce Chemist
   Laboratory
   Tanteen
   ST. GEORGE'S
   Tel: (809) 440 3273
   Fax: (809) 440 4191

3. The Secretary
   Pesticide Control Board
   c/o Produce Chemist
   Laboratory
   Tanteen
   ST. GEORGE'S
   Tel: 440 3273
   Fax: 440 4191

4. Pest Management Unit
   c/o The Permanent Secretary
   Ministry of Agriculture
   ST. GEORGE'S
   Tel: 440 2248
   Fax: 440 4191

5. Environmental Health
   Department
   c/o Permanent Secretary
   Ministry of Health
   The Carenage
   ST. GEORGE's
   Tel: 440 2846
   Fax: 440 4127

6. Mr S. Breo
   Ministry of Health
   ST. GEORGE'S
   Tel: (1809) 440 3485
   Fax: (1809) 440 4127

Guatemala
República de Guatemala

Responsibility	
Main authority responsible for approval/registration of pesticides	1
Main authority responsible for pesticide use directives, permits and training	
Main authority responsible for establishing national maximum pesticide residue limits	2
Main authority responsible for control of pesticide residues in food	3, 10, 11
Main laboratory for analysis of pesticide residues in food	3
Main laboratory for quality control of pesticide products	
Main organisation responsible for applied research regarding pesticide use. Information on related regional research	3, 4
Designated national authority for notifications and information regarding banned and severely restricted pesticides including the PIC procedure	5
Codex Alimentarius contact point	6
National IRPTC correspondent	7
Infoterra national focal point	8
National IFCS focal point	-
National association of pesticide manufacturers	9, 12, 13

Note

Poisons information/control centre: Departamento de Toxicologia, Faculdad de CC.QQ. y Farmacia, Universidad de San Carlos, 3 Calle 6-47, Zona 1, Guatemala C.A.

Responsible organisations

1. Departamento de Control y Registro de Agroquímicos
   7a Avenida 3-14 Zona 13
   CIUDAD DE GUATEMALA

2. Comisión Guatemalteca de Normas (COGUANOR)
   6a Avenida 11-43 Zona 1
   20 Nivel, Local C
   CIUDAD DE GUATEMALA

3. Laboratorio Unificado de Control de Alimentos y Medicamentos (LUCAM)
   Aparado 1188
   CIUDAD DE GUATEMALA

4. Instituto de Ciencias Tecnología Agricola (ICTA)
   Avenida reforma 8-60 Zona 9
   CIUDAD DE GUATEMALA

5. Jefe,
   Departamento de Control y Registro de Agroquímicos
   Dirección Técnica de Sanidad Vegetal
   7a. Avenida 3-67, Zona 13
   CIUDAD DE GUATEMALA
   Tel: (502 2) 720 163, 720 432
   Fax: (502 2) 720 163

6. Dr L R Sandoval Cambara
   Director-Técnico,
   Inspección Sanitaria y Control de Alimentos de Origen Animal-DIGESEPE
   Ministerio de Agricultura, Ganadería y Alimentación
   Palacio Nacional
   CIUDAD DE GUATEMALA

7. Comisión Asesora de la Comisión Ministerial para la Conservación Mejoramiento del Medio Ambiente
   Ministerio de Gobernación
   9a Avenida Entre 14 y 15 Calles, Zona 1
   CIUDAD DE GUATEMALA

8. Sra Dora Luz Mazariegos de Perez
   Comisión Nacional del Medio Ambiente (CONAMA)
   7a Avenida 4-35, Zona 1
   CIUDAD DE GUATEMALA
   Tel: (502 2) 21816
   Fax: (502 2) 535 109

9. AGREQUIMA/FECCOPIA
   5A Avenida 15-45, Zona 10
   Centro Empresarial
   Torre 1, 9 Nivel
   CIUDAD DE GUATEMALA
   Tel: (502 2) 337 082
   Fax: (502 2) 337 081

10. Dirección Técnica de Inspección Sanitaria y Control de Alimentos de Origen Animal para Consumo Humano
    Avenida reforma 8-60 Zona 9
    GIUDAD DE GUATEMALA

11. Departamento de Control
    10 Avenida 14-65 Zona 1
    CIUDAD DE GUATEMALA

12. Federation Centro-Americana y del Caribe de Associaciones Gremiales y Camaras Provedores de Productos Agropecuarios,
Blvd. Vista Hermosa 22-65,
Zona 15, Commercial Metro 15,
3er Nivel, Local 11
01015 CIUDAD DE GUATEMALA

13. Germail de Proveedores de Productos Para la Agricultura
Calle 7-21, Zona 9
CIUDAD DE GUATEMALA

Guinea

République de Guinée

Responsibility	
Main authority responsible for approval/registration of pesticides	
Main authority responsible for pesticide use directives, permits and training	
Main authority responsible for establishing national maximum pesticide residue limits	
Main authority responsible for control of pesticide residues in food	
Main laboratory for analysis of pesticide residues in food	
Main laboratory for quality control of pesticide products	
Main organisation responsible for applied research regarding pesticide use. Information on related regional research	
Designated national authority for notifications and information regarding banned and severely restricted pesticides including the PIC procedure	1, 2
Codex Alimentarius contact point	4
National IRPTC correspondent	3
Infoterra national focal point	5
National IFCS focal point	6
National association of pesticide manufacturers	

Responsible organisations

1. Head,
   Division prevention de la pollution et controle des nuisances
   Direction nationale de l'environnement
   B.P. 3118
   CONAKRY
   Fax: 415 004

2. Monsieur le Directeur,
   Division de la protection des végétaux
   Direction nationale de l'agriculture
   B.P. 576
   CONAKRY

3. Monsieur le Chef
   Section controle des produits chimique
   Correspondant national du RISCPT
   Direction nationale de l'environnement
   B.P. 3118
   CONAKRY

4. M. le Directeur
   Institut de Normalisation et de Métrologie
   c/o Ministère de l'Industrie, du Commerce et de l'Artisanat
   B.P. 468
   CONAKRY

5. Mme. Aissatou Bah
   PFN d'Infoterra
   Direction nationale de l'environnement
   Ministère des ressources naturelles et de l'environnement
   B.P. 3118
   CONAKRY
   Tel: (224) 443 742, 443 868

6. Mr L Conte
   Chef,
   Etablissements Classes
   Direction Nationale de l'Environnement
   Ministere de l'energie et de l'environnement
   B.P.3302
   CONAKRY
   Tel: (224) 443 868, 443 742
   Fax: (224) 414 839

Guinea – Bissau

República da Guiné-Bissau

Responsibility	
Main authority responsible for approval/registration of pesticides	1
Main authority responsible for pesticide use directives, permits and training	
Main authority responsible for establishing national maximum pesticide residue limits	-
Main authority responsible for control of pesticide residues in food	1, 2, 3
Main laboratory for analysis of pesticide residues in food	-
Main laboratory for quality control of pesticide products	
Main organisation responsible for applied research regarding pesticide use. Information on related regional research	-
Designated national authority for notifications and information regarding banned and severely restricted pesticides including the PIC procedure	-
Codex Alimentarius contact point	1
National IRPTC correspondent	-
Infoterra national focal point	4
National IFCS focal point	-
National association of pesticide manufacturers	-

Responsible organisations

1. Ministére de developpement rural et de l'agriculture
   Rural e Pescas
   C.P. 71
   1011 BISSAU

2. Laboratório dos Servicos de Veterinaria
   C.P. 26
   BISSAU

3. Laboratório Nacional de Saude Pública
   C.P. 50
   BISSAU

4. Mr. Cipriano Cassama
   PFN d'Infoterra
   Direction des Forêts et Chasse
   Ministère du Développement Rural et de l'Agriculture
   C.P. 71
   BISSAU
   Tel: (245) 215474

Guyana
Republic of Guyana

Responsibility	
Main authority responsible for approval/registration of pesticides	-
Main authority responsible for pesticide use directives, permits and training	
Main authority responsible for establishing national maximum pesticide residue limits	-
Main authority responsible for control of pesticide residues in food	1
Main laboratory for analysis of pesticide residues in food	1
Main laboratory for quality control of pesticide products	
Main organisation responsible for applied research regarding pesticide use. Information on related regional research	2
Designated national authority for notifications and information regarding banned and severely restricted pesticides including the PIC procedure	-
Codex Alimentarius contact point	3
National IRPTC correspondent	4
Infoterra national focal point	5
National IFCS focal point	-
National association of pesticide manufacturers	-

Responsible organisations

1. Government Analyst Department
   19-20 Evans & Lyng Streets
   Charlestown
   GEORGETOWN
   Tel: 63711, 68337
   Fax: (592 2) 257524

2. National Agricultural Research Institute
   East Coast Demerara
   MON REPOS
   Tel: 020-2842

3. Dr. C Ramcharran
   Director,
   Guyana National Bureau of Standards
   P.O. Box 10926
   GEORGETOWN
   Tel: 56226, 57455, 59013

4. The Head,
   Plant Protection
   National Agricultural Research Institute
   East Coast Demerara
   MON REPOS
   Tel: 020-2881

5. Ms Mavis Taylor,
   Infoterra NFP Officer
   c/o UNDP
   P.O. Box 10960
   GEORGETOWN
   Telex: 2201 gy
   Fax: (592 2) 62942

Haiti

République d'Haiti, Repiblik Dayti

Responsibility	
Main authority responsible for approva/registration of pesticides	1
Main authority responsible for pesticide use directives, permits and training	
Main authority responsible for establishing national maximum pesticide residue limits	1
Main authority responsible for control of pesticide residues in food	2
Main laboratory for analysis of pesticide residues in food	-
Main laboratory for quality control of pesticide products	
Main organisation responsible for applied research regarding pesticide use. Information on related regional research	3
Designated national authority for notifications and information regarding banned and severely restricted pesticides including the PIC procedure	7
Codex Alimentarius contact point	4
National IRPTC correspondent	5
Infoterra national focal point	6
National IFCS focal point	-
National association of pesticide manufacturers	-

Responsible organisations

1. Ministry of Agriculture,
   Natural Resources and
   Rural Development
   DAMIEN

2. Ministry of Commerce and
   Industry
   Champ-de-Mars
   PORT-AU-PRINCE

3. Centre for Agricultural
   Research and
   Documentation
   DAIMEN

4. M. Raymond Tardieu
   Direction Normalisation et
   Contrôle de la Qualité
   Ministère du Commerce
   8 Rue Légitime
   Champ-de-Mars
   PORT-AU-PRINCE

5. M. le Directeur
   Département de la Santé
   Publique et de la Population
   c/o Service de Contrôle des
   Medicaments, des
   Pharmacies et des
   Narcotiques
   PORT-AU-PRINCE

6. Mme. Dalberg Claude
   PFN d'Infoterra
   Chef,
   Service de protection de
   l'environnement
   Ministère de l'agriculture et
   des resources naturelles et
   du développement rural
   P.O. Box 1441
   PORT-AU-PRINCE
   Tel: (509 1) 21867
   Fax: (509 1) 23599

7. Service de la protection des
   vegetaux
   Unite'de protection sanitaire/
   MARNDR
   Route nationale No.1
   DAMIEN
   Tel: 228 637
   Fax: 223 595

Honduras

República de Honduras

Responsibility	
Main authority responsible for approval/registration of pesticides	1
Main authority responsible for pesticide use directives, permits and training	
Main authority responsible for establishing national maximum pesticide residue limits	-
Main authority responsible for control of pesticide residues in food	2
Main laboratory for analysis of pesticide residues in food	-
Main laboratory for quality control of pesticide products	
Main organisation responsible for applied research regarding pesticide use. Information on related regional research	3
Designated national authority for notifications and information regarding banned and severely restricted pesticides including the PIC procedure	4
Codex Alimentarius contact point	5
National IRPTC correspondent	-
Infoterra national focal point	6
National IFCS focal point	-
National association of pesticide manufacturers	7

Responsible organisations

1. Ministerio de Recursos Naturales
Programa de Sanidad Vegetal
TEGUCIGALPA, D.C.

2. Federación Hondurena de Investigación Agricola (FHIA)
LA LIMA (PROGRESO)

3. Secretaria de Recursos Naturales
TEGUCIGALPA, D.C.

4. Jefe del Departamento de Sanidad Vegetal
Boulevard Centro América
Ave. FAO
TEGUCIGALPA
Fax: (504) 326 213

5. Dra. Georgina Nazar
Division de Control de Alimentos
Dirección General de Salud
Edificio CESCO-ALIMENTOS
Barrio Morazan
TEGUCIGALPA, M.D.C.
Tel: (504) 321 139
Fax: (504) 312 713

6. Dr. Carlos A. Medina
CNC de Infoterra
Secretaria del Ambiente
Calle La Fuente No. 1501
Apartado Postal 4710
TEGUCICALPA
Tel: (504) 375 664
Fax: (504) 375 726

7. Asociacion de Distribuidores, Importadores y Vendedores de Productos Agropecuarios de Honduras (ADIVEPAH)
Apartado Postal 3642
COMAYAGÜELA
Tel: (504) 333 662
Fax: (504) 320 359

Hungary
Magyar Köztársaság

Responsibility

Main authority responsible for approval/registration of pesticides	1
Main authority responsible for pesticide use directives, permits and training	
Main authority responsible for establishing national maximum pesticide residue limits	2, 3
Main authority responsible for control of pesticide residues in food	1, 4, 5, 6
Main laboratory for analysis of pesticide residues in food	4, 5, 6
Main laboratory for quality control of pesticide products	
Main organisation responsible for applied research regarding pesticide use. Information on related regional research	1, 4, 6, 7, 8, 9
Designated national authority for notifications and information regarding banned and severely restricted pesticides including the PIC procedure	10
Codex Alimentarius contact point	11
National IRPTC correspondent	7
Infoterra national focal point	12
National IFCS focal point	14
National association of pesticide manufacturers	13

Note

Poisons information/control centre: Egeszsegugyi Toxikologiai Infomacios Szolgalat, Orszagos Munka es Uzemegeszsegugyi, Postafiok 22, Nagyvarad ter 2, 1450 Budapest, Tel: (36 1) 113 7890, Fax: (36 1) 113 6891 [Head: Dr. Z. Adamis]

Responsible organisations

1. MÉM Növényvédelmi és Agrokémiai Föosztály
P.O. Box 55
H-1860 BUDAPEST

2. Állami Közegészségügyi és Járványügyi Föfelügyelöség, Egészségügyi Minisztérium
P.O. Box 251
H-1502 BUDAPEST

3. MÉM Állategészségügyi és Élelmiszerhigiéniai Föosztály
P.O. Box 55
H-1860 BUDAPEST

4. Országos Élelmezés-és Táplálkozás-tudományi Intézet
(OETI)
P.O. Box 52
H-1476 BUDAPEST

5. Allategészségügyi és Élelmiszerellenörzö Központ
P.O. Box 13
H-1453 BUDAPEST
Tel: 135 440

6. MÉM Növényvédelmi és Agrokémiai Központ (NAK)
P.O. Box 127
H-1502 BUDAPEST
Tel: 851 117

7. The Director
National institute of Occupational Health
P.O. Box 22
H-1450 BUDAPEST 9
Tel: (361) 215 54 91
Fax: (361) 215 68 91

8. MTA Növényvédelmi Kutatóintézet
P.O. Box 102
H-1525 BUDAPEST

9. Országos Közegészségügyi Intézet
Egészégügyi Minisztérium
P.O. Box 64
H-1966 BUDAPEST

10. The Director
Plant Health and Soil Conservation Department
Ministry of Agriculture and Food
Kossuth L. tér 11
H-1055 BUDAPEST
Tel: (36 1) 153 3000
Fax: (36 1) 153 0518

11. Mr I Olah
Secretary,
Hungarian National Codex Committee
Hungarian Office for Standardization
P.O. Box 24
H-1450 BUDAPEST 9
Tel: (36 1) 218 3011
Fax: (36 1) 218 5125

12. Mr. Endredy Istvá
Infoterra NFP
Környezetgazdalkodasi Intezet (KGI)
V., Alkotmamy U. 29
P.O. Box 1369
Pf.352
BUDAPEST 5
Tel: (36 1) 111 8278, 132 2331
Fax: (36 1) 111 5826, 112 9872

Responsible organisations

13. Hungarian Chemical Industry Association (HCIA)
Erzsebet Királyne
Utja 1/c
PF40
H-1406 BUDAPEST 76
Tel: (36 1) 153 1933
Fax: (36 1) 142 0980

14. Deputy Secretary of State
National Institute of Occupational Health
Attn: Dr E Podmaniczku
Ministry of Welfare
Department of International Relations
Arany Janos 6-8
H-1361 BUDAPEST
Tel: (36 1) 132 3100
Fax: (36 1) 111 8054

Iceland
Lydveldid Island

Responsibility	
Main authority responsible for approval/registration of pesticides	1, 2
Main authority responsible for pesticide use directives, permits and training	2, 6
Main authority responsible for establishing national maximum pesticide residue limits	1, 2, 3, 4
Main authority responsible for control of pesticide residues in food	3, 4
Main laboratory for analysis of pesticide residues in food	3, 5
Main laboratory for quality control of pesticide products	
Main organisation responsible for applied research regarding pesticide use. Information on related regional research	6
Designated national authority for notifications and information regarding banned and severely restricted pesticides including the PIC procedure	10
Codex Alimentarius contact point	8
National IRPTC correspondent	7
Infoterra national focal point	9
National IFCS focal point	7
National association of pesticide manufacturers	-

Responsible organisations

1. Umhverfisraduneytid
Vonarstraeti 4
IS-150 REYKJAVIK
Tel: (354) 560 9600
Fax: (354) 562 4566

2. Hollustuvernd rikisins
eiturefnasvid
P.O. Box 8080
IS-128 REYKJAVIK
Tel: (354) 568 8848
Fax: (354) 568 1896

3. Hollustuvernd ríkisins
Armula 1a
P.O. Box 8080
IS-128 REYKJAVIK
Tel: (354) 568 8848
Fax: (354) 568 1896

4. Manneldisrád
Armula 1a
P.O. Box 8080
IS-125 REYKJAVIK
Tel: (354) 568 8848
Fax: (354) 568 1896

5. Rannsóknastofnun
Fiskidnadarins
Skúlagötu 4
IS-101 REYKJAVIK
Tel: (354) 562 0240
Fax: (354) 562 0740

6. Rannsoknastofnun
Landbunadarins
Keldnaholt
IS-112 REYKJAVIK
Tel: (354) 577 1010
Fax: (354) 577 1030

7. Mrs. S Gisladottir
Director of Office,
Office of Chemicals
Environmental and Food
Agency
P.O. Box 8080
IS-128 REYKJAVIK
Tel: (354) 568 8848
Fax: (354) 568 1896

8. Mr. J. Gislason,
Director of Office
Environmental and Food
Agency
Office of Food and Hygiene
P.O. Box 8080
IS-128 REYKJAVIK
Tel: (354) 568 8848
Fax: (354) 568 1896

9. Mr. Jon Erlendsson
Director,
Scientific and Technical
Information Service
University of Iceland
P.O. Box 7066
IS-127 REYKJAVIK
Tel: (354) 525 4665

10. Director,
Environmental and Food
Agency
Armula 1
P.O. Box 8080
IS-128 REYKJAVIK
Tel: (354) 568 8848
Fax: (354) 568 1896

India

Bharat, Republic of India

Responsibility	
Main authority responsible for approval/registration of pesticides	1
Main authority responsible for pesticide use directives, permits and training	
Main authority responsible for establishing national maximum pesticide residue limits	2
Main authority responsible for control of pesticide residues in food	1, 3
Main laboratory for analysis of pesticide residues in food	2, 4, 5, 6, 7, 8, 9
Main laboratory for quality control of pesticide products	
Main organisation responsible for applied research regarding pesticide use. Information on related regional research	14
Designated national authority for notifications and information regarding banned and severely restricted pesticides including the PIC procedure	15
Codex Alimentarius contact point	16
National IRPTC correspondent	12
Infoterra national focal point	13
National IFCS focal point	-
National association of pesticide manufacturers	10, 11, 17

Responsible organisations

1. Ministry of Agriculture and Rural Development
   Department of Agriculture and Rural Development
   Krishi Bhavan
   NEW DELHI

2. Ministry of Health and Family Welfare
   Nirman Bhavan
   NEW DELHI

3. Assistant Director-General (PFA)
   Directorate General of Health Services
   Nirman Bhavan
   NEW DELHI

4. Central Food Laboratory
   3 Kyd Street
   CALCUTTA 16

5. Central Food Laboratory
   c/o Central Food Technological Research Institute
   MYSORE 570013

6. Central Food Laboratory
   c/o Food Research Standardisation Laboratory
   Navyug Market
   GHAZIABAD

7. Central Food Laboratory
   c/o State Public Health Laboratory
   PUNE 411001

8. Central Insecticide Laboratories
   Directorate of Plant Protection
   Ministry of Rural Reconstruction
   NH.IV
   FARIDABAD

9. Bureau of Indian Standard
   Manak Bhavan
   9 BS Zafar Marg
   NEW DELHI

10. Pesticides Association of India (PAI)
    1202 New Delhi House
    27 Barakhamba Road
    NEW DELHI 110001
    Tel: (91 11) 331 5776, 331 3179
    Fax: (91 11) 331 3179

11. Association of Basic Manufacturers of Pesticides
    c/o BASF India Ltd.
    Rhone-Poulenc House
    BOMBAY 400 025
    Tel: (91 22) 493 0703, 636 0866
    Fax: (91 22) 634 9590

12. The Director
    Industrial Toxicology Research Centre
    Mahatma Gandhi Marg
    P.O. Box 80
    LUCKNOW 226001 U.P.

13. Mr Harjit Singh
Infoterra NFP
Adviser (EI-Division)
Ministry of Environment and Forests
B Block, Paryavaran Bhavan
CGO Complex
Lodhi Road
NEW DELHI 110003
Tel: (91 11) 4360 419
Fax: (91 11) 436 1228

14. The Plant Protection Adviser to the Government of India
Department of Plant Protection, Quarantine and Storage
NH-VI
FARIDABAD

15. The Director/Deputy Secretary
Plant Protection Division
Department of Agriculture and Co-operation
Room No. 244-A
Krishi Bhavan
NEW DELHI

16. Ms. Debi Mukherjee
Assistant Director General
Central Committee for Food Standards
National Codex Committee
Directorate General of Health Services
Nirman Bhavan
NEW DELHI 110011
Tel: 3012290

17. Pesticides Formulators of India (PFAI)
B-4, Anand Co-op. Housing Society
Sitladevi Temple Road
Mahim (West)
BOMBAY 400 016
Tel: (91 22) 437 5279
Fax: (91 22) 611 6736

Indonesia
Republik Indonesia

Responsibility	
Main authority responsible for approval/registration of pesticides	1
Main authority responsible for pesticide use directives, permits and training	
Main authority responsible for establishing national maximum pesticide residue limits	2, 3, 6
Main authority responsible for control of pesticide residues in food	1, 4
Main laboratory for analysis of pesticide residues in food	4, 5
Main laboratory for quality control of pesticide products	
Main organisation responsible for applied research regarding pesticide use. Information on related regional research	1, 6
Designated national authority for notifications and information regarding banned and severely restricted pesticides including the PIC procedure	7, 8
Codex Alimentarius contact point	9
National IRPTC correspondent	10
Infoterra national focal point	11
National IFCS focal point	13
National association of pesticide manufacturers	12

Note

Poisons information/control centre: Unit Informasi Bahan Berbahaya, Ditwas Narkoba, Ditjen POM, Ministry of Health, Jln. Percetakkan Negara 23, Jakarta 10560, Tel: (62) 21 41 55 23, FaX: (62) 21 41 55 23 [Head: Dr (Ms) M Z Hanafy]

Responsible organisations

1. Direktorat Jenderal Perlindungan
   Tanaman Pangan
   Departemen Pertanian
   Jl. Harsono R M, Ragunan
   Pasar Minggu
   JAKARTA

2. Departemen Kesehatan
   Jl. Prapatan No 10
   JAKARTA PUSAT

3. Direktorat Narkoba
   Direktorat Jenderal POM
   Jl. Percetakan Negara No 23
   JAKARTA

4. Pusat Pemeriksaan Obat dan Makanan
   Jl. Percetakan Negara No 23
   JAKARTA

5. Pusat Penelitian Farmasi
   Jl. Percetakan Negara No 29
   JAKARTA

6. Badan Penelitian dan Pengembangan
   Deparemen Pertanian Ragunan
   Pasar Minggu
   JAKARTA

7. Ms Masnellyarti Hilman
   Bapedal Offices
   Arthaloka Building, 11th Floor
   Jl. Jend. Sudirman No. 2
   JAKARTA PUSAT
   Tel: (62 21) 583 918
   Fax: (62 21) 570 3365

8. The Chairman,
   Pesticides Committee
   Direktorat Bina Perlindungan Tanaman
   Jln. AUP. Pasar Minggu
   JARKARTA SELATAN 12520
   Tel: (62 21) 780 5652, 780 6213

9. M Bambang H Hadiwiardjo
   Dewan Standardisasi Nasional (DSN)
   Gedung PDIN-LIPI
   Jl. Jen. Gatot Subroto
   P.O. Box 3123
   12190 JAKARTA

10. The Chairman
    Technical Team for Toxic and Hazardous Substances Management
    State Ministry for Population and Environment
    Jalan Merdeka Barat 15
    JAKARTA

11. Ms Utari Budihardjo
    Head, Dissemination of Scientific Information Division
    Centre for Scientific Documentation and Information
    PDII-LIPI
    P.O. Box 4298
    12042 JAKARTA
    Tel: (62 21) 5733465, 5251063
    Fax: (62 21) 5733467

Responsible organisations

12. Association of Indonesian
    Pesticide Industries
    Perdatam Raya No. 9
    JAKARTA 12780
    Tel: (62 21) 714 845,
    799 8832
    Fax: (62 21) 770 400,
    799 4252

13. Mr. H. Abdullah Nawawi
    Director,
    Directorate of Narcotics
    Control and Other
    Dangerous Substances
    Department of Health
    Jl. Perctakan Negara No. 23
    JAKARTA 10560
    Tel: (62 21) 420 7683
    Fax: (62 21) 420 7683

Iran

Jomhuri-ye Eslami-ye Iran, Islamic Republic of Iran

Responsibility

Main authority responsible for approval/registration of pesticides	1
Main authority responsible for pesticide use directives, permits and training	2
Main authority responsible for establishing national maximum pesticide residue limits	3
Main authority responsible for control of pesticide residues in food	4
Main laboratory for analysis of pesticide residues in food	4
Main laboratory for quality control of pesticide products	2
Main organisation responsible for applied research regarding pesticide use. Information on related regional research	3
Designated national authority for notifications and information regarding banned and severely restricted pesticides including the PIC procedure	7
Codex Alimentarius contact point	6
National IRPTC correspondent	5
Infoterra national focal point	8
National IFCS focal point	-
National association of pesticide manufacturers	9

Note

Poisons information/control centre: Imam Reza Hospital, P.O. Box 348, Mashhad 91735, Tel: (98 51) 93043, Fax: (98 51) 47254 [Head: Pr. M. Balali-Mood]

Responsible organisations

1. Pesticide Supervision Committee
   Plant Protection Organization
   Ministry of Agriculture
   P.O. Box 19395-4568
   TEHRAN
   Tel: (98 21) 240 2046
   Fax: (98 21) 240 3197

2. Plant Protection Organization
   Ministry of Agriculture
   P.O. Box 19395-4568
   TEHRAN
   Tel: (98 21) 240 2046
   Fax: (98 21) 240 3197

3. Plant Pests and Diseases Research Institute
   Ministry of Agriculture
   P.O. Box 19395-1454
   TEHRAN
   Tel: (98 21) 240 3012
   Fax: (98 21) 240 3691

4. Food and Drug Control Laboratory
   Ministry of Health and Medical Education
   Imam Khomeini Avenue 31
   11136 TEHRAN
   Tel: (98 21) 640 081,
   640 7556
   Fax: (98 21) 640 4330

5. The Director General
   Environmental Health Department
   Ministry of Health and Medical Education
   Attn: Mr B Hosseini
   TEHRAN
   Tel: (98 21) 677 682
   Telex: 213423 ihrd

6. Instutute of Standards and Industrial Research
   Ministry of Industries
   P.O. Box 15875-4618
   TEHRAN
   Tel: (98 21) 226 031
   Fax: (98 21) 880 2276

7. The Secretary,
   High Council of Policy and Planning for Reduction of Agricultural Pesticides
   Center for Agricultural Planning and Economic Studies (CAPES)
   Ministry of Agriculture
   Attn: Mr. Abdovais Pourmand
   P.O. Box 15815-1515
   TEHRAN
   Tel: (98 21) 833 1769
   Fax: (98 21) 882 9459

8. Infoterra NFP
   Advisor, Chief of the Cabinet of the Vice-President, and Director-General of International Affairs
   Department of Environment
   P.O. Box 5181
   15875 TEHRAN
   Tel: (98 21) 883 1769
   Fax: (98 21) 891 005, 698 246

9. Fertilizer Distrubution and Pesticides Procurement Organization
   73, Zartosht Avenue
   TEHRAN
   Tel: (98 21) 655 667

Iraq
Al Jumhuriyah al-Iraqiyah

Responsibility	
Main authority responsible for approval/registration of pesticides	1, 2
Main authority responsible for pesticide use directives, permits and training	
Main authority responsible for establishing national maximum pesticide residue limits	3
Main authority responsible for control of pesticide residues in food	3, 4
Main laboratory for analysis of pesticide residues in food	3, 4
Main laboratory for quality control of pesticide products	
Main organisation responsible for applied research regarding pesticide use. Information on related regional research	5
Designated national authority for notifications and information regarding banned and severely restricted pesticides including the PIC procedure	8
Codex Alimentarius contact point	3
National IRPTC correspondent	6
Infoterra national focal point	7
National IFCS focal point	-
National association of pesticide manufacturers	-

Note

Poisons information/control centre: University of Baghdad, Baghdad [Head Dr. A. Khaladi]

Responsible organisations

1. Ministry of Health
   P.O Box 2130
   BAGHDAD
   Tel: 718 1970

2. Directorate-General for Plant Protection
   P.O. Box 1187
   Simak
   Sadoun Street
   BAGHDAD
   Tel: 887 3011

3. Central Organization for Standardization and Quality Control
   Ministry of Planning
   P.O. Box 13032
   Aljadiria
   BAGHDAD
   Tel: 776 5180
   Fax: 776 5781

4. Nutrition Research Institute
   P.O. Box 4010
   Al-Adhamya
   BAGHDAD
   Tel: 422 6015

5. Biological Research Centre
   Environmental Pollution Department
   Jadiriya
   P.O. Box 2371
   BAGHDAD
   Tel: 776 5116

6. The Director General
   Directorate General for Human Environment
   Aquba Bin Nafia Square
   BAGHDAD
   Tel: 98827

7. Dr. Abdul Jabar Abdul Abbas
   Director,
   Preventive Medicine and Environment Protection Directorate
   P.O. Box 10062
   BAGHDAD
   Tel: (964 1) 719 2033

8. General Authority for Plant Protection
   Ministry of Agriculture
   BAGHDAD

Ireland

Eire

Responsibility	
Main authority responsible for approval/registration of pesticides	1
Main authority responsible for pesticide use directives, permits and training	1, 8
Main authority responsible for establishing national maximum pesticide residue limits	1
Main authority responsible for control of pesticide residues in food	1, 2, 3, 4
Main laboratory for analysis of pesticide residues in food	1, 5
Main laboratory for quality control of pesticide products	1
Main organisation responsible for applied research regarding pesticide use. Information on related regional research	7, 8, 9
Designated national authority for notifications and information regarding banned and severely restricted pesticides including the PIC procedure	1
Codex Alimentarius contact point	10
National IRPTC correspondent	11
Infoterra national focal point	12
National IFCS focal point	13
National association of pesticide manufacturers	6

Note

Poisons information/control centre: Beaumont Hospital, P.O. Box 1297, Dublin 9, Tel: (353 1) 379964, Fax: (353 1) 376982 [Head: Dr. J. A. Tracey]

Responsible organisations

1. Pesticide Control Service
   Department of Agriculture,
   Food and Forestry
   Abbotstown Laboratory
   Complex
   Abbotstown, Castleknock
   DUBLIN 15
   Tel: (353 1) 607 2613
   Fax: (353 1) 820 4260

2. Public Analysts Laboratory
   GALWAY
   Co Galway
   Tel: 64448

3. Public Analysts Laboratory
   CORK
   Co Cork
   Tel: 966555

4. Public Analysts Laboratory
   Corn Market
   DUBLIN 8
   Tel: 776450

5. State Laboratory
   Abbotstown
   CASTLEKNOCK
   Co Dublin
   Tel: 217853
   Telex: 33735

6. Animal and Plant Health
   Association (APHA)
   Franklin House
   140 Pembroke Road
   DUBLIN 4
   Tel: (353 1) 660 3350
   Fax: (353 1) 668 6672

7 Ieagase Agricultural Institute
   KINSEALY
   Co Dublin
   Tel: 460644
   Telex: 31479

8. Ieagase Agricultural Institute
   Oakpark
   CARLOW
   Tel: 31425
   Telex: 33038

9. Department of Agriculture,
   Food and Forestry
   Backweston
   LEIXLIP
   Co Kildare
   Tel: 628 0608

10. The Secretary
    Irish National FAO
    Committee
    Department of Agriculture and
    Fisheries
    Agriculture House
    DUBLIN 2
    Tel: 789011
    Fax: 661 6263

11. The Manager
    Environmental Services
    Institute for Industrial
    Research and Standards
    (IIRS)
    Shannon Town Centre
    Co Clare
    SHANNON

12. Mr Noel Hughes
Infoterra NFP
The Environmental
Information Service
17, St Andrew Street
DUBLIN 2
Tel: (353 1) 6793144
Fax: (353 1) 6795204

13. Department of Foreign Affairs
DUBLIN 2
Tel: (353 1) 732 8550
Fax: (353 1) 732 8106,
731 4365

Israel

Medinat Yisrael, Israil

Responsibility	
Main authority responsible for approval/registration of pesticides	1
Main authority responsible for pesticide use directives, permits and training	
Main authority responsible for establishing national maximum pesticide residue limits	3
Main authority responsible for control of pesticide residues in food	3, 4
Main laboratory for analysis of pesticide residues in food	4
Main laboratory for quality control of pesticide products	
Main organisation responsible for applied research regarding pesticide use. Information on related regional research	5
Designated national authority for notifications and information regarding banned and severely restricted pesticides including the PIC procedure	1, 6
Codex Alimentarius contact point	7
National IRPTC correspondent	8
Infoterra national focal point	9
National IFCS focal point	2
National association of pesticide manufacturers	10, 11

Note

Poisons information/control centre: Rambam Medical Centre. P.O. Box 9602, Haifa 31096, Fax: (972 4) 514749 [Head: Dr. U. Taitelman]

Responsible organisations

1. Department of Plant Protection and Inspection
   Ministry of Agriculture
   P.O. Box 78
   BET-DAGAN 50250
   Tel: (972 3) 968 1500
   Fax: (972 3) 968 1507

2. Dr. M S Tirosh
   Director,
   Clinical Toxicology Unit
   Chaim Sheba Medical Centre
   TEL-HASHOMER 52621
   Tel: (972 3) 534 9512
   Fax: (972 3) 530 3366

3. Environmental Health Unit
   Ministry of Health
   P.O. Box 1176
   JERUSALEM 91010
   Tel: (2) 247 471

4. Institute for the Standardization and Control of Pharmaceuticals
   P.O. Box 1457
   JERUSALEM 91010
   Tel: (2) 227 402

5. Division of Pesticide Residues
   Plant Protection Institute
   Volcani Center
   P.O. Box 6
   BET DAGAN 50250
   Tel: (3) 480 440

6. The Head,
   Hazardous Substances Department
   Ministry of the Environment
   P.O. Box 34033
   JERUSALEM 95464
   Tel: (972 2) 655 3765
   Fax: (972 2) 655 3763

7. Mr Dan Halpern
   Israel Codex Alimentarius Committee
   Ministry of Industry and Trade
   P.O. Box 299
   JERUSALEM 91002
   Tel: (972 3) 566 0004
   Fax: (972 3) 560 5146

8. The Director-General
   Ministry of the Environment
   Attn: Dr. I Peleg
   P.O. Box 34033
   JERUSALEM 95464

9. Ms. Esther Korn-Ancri
   Infoterra NFP
   Ministry of the Environment
   P.O. Box 6234
   JERUSALEM 91061
   Tel: (972 2) 701 604
   Fax: (972 2) 513 945

10. Pesticide Manufacturers Association of Israel (PMAI)
    c/o Makheshim Agan International Coordination Centre
    P.O. Box 7
    1050 BRUSSELS
    BELGIUM
    Tel: (32 2) 646 8606
    Fax: (32 2) 646 9152

11. Pesticides Manufacturers Association in Israel
    c/o Dead Sea Bromine Co. Ltd.
    P.O. Box 180
    BEERSHEVA 84101
    Tel: (972 57) 667 260

Italy

Repubblica Italiana

Responsibility	
Main authority responsible for approval/registration of pesticides	1
Main authority responsible for pesticide use directives, permits and training	
Main authority responsible for establishing national maximum pesticide residue limits	1
Main authority responsible for control of pesticide residues in food	2, 3
Main laboratory for analysis of pesticide residues in food	-
Main laboratory for quality control of pesticide products	
Main organisation responsible for applied research regarding pesticide use. Information on related regional research	4, 5, 6
Designated national authority for notifications and information regarding banned and severely restricted pesticides including the PIC procedure	7
Codex Alimentarius contact point	8
National IRPTC correspondent	9
Infoterra national focal point	10
National IFCS focal point	-
National association of pesticide manufacturers	11

Note

Poisons information/control centres in Bologna, Catania, Cesena, Chieti, Genova, La Spezia, Lecce, Milan, Napoli, Pordenone, Reggio Calabria, Roma and Trieste

Responsible organisations

1. Ministero della Sanità
   Direzione Generale per l'Igiene degli Alimenti e Nutrizione
   Piazza G Marconi 25
   I-00144 ROMA
   Tel: (6)5994
   Fax: (6)5925936

2. Istituto Nazionale della Nutrizione
   Via Adreatina 546
   I-00179 ROMA
   Tel: (6)5032412
   Fax: (6)5031592

3. Istituto Superiore di Sanità (ISS)
   Viale Regina Elena 299
   I-00161 ROMA
   Tel: (6)4990
   Fax: (6)4469938

4. Osservatorio per le Malattie delle Piante Regione Liguria
   Via Nino Bixio 6
   I-161 28 GENOVA
   Tel: (10)589729
   Fax: (10)5484909

5. Ministero dell'Agricoltura e delle foreste
   Direzione Generale della Produzione Agricola-Fitopatologia
   Via XX Settembre 20
   I-001 ROMA
   Tel: (6)4665
   Fax: (6)483998

6. Osservatorio per le Malattie delle Piante
   Regione Emilia-Romagna
   Via di Corticella 133
   I-40129 BOLOGNA
   Tel: (51)352917
   Fax: (51)352138

7. Dott.ssa Fernanda Ferrazzin
   Ministero della Sanita
   Direzione Generale Igiene Alimenti e Nutrizione,
   Div. III
   Piazza Marconi 25
   I-00144 ROMA
   Tel: (39 6) 599 43844
   Fax: (39 6) 599 43598

8. Comitato Nazionale Italiano per il Codex Alimentarius
   Direzione Generale della Tutela Economica dei Prodotti Agricoli
   Via Sallustiana 10
   I-00187 ROMA
   Tel: (6) 488 1252
   Fax: (6) 474 3971

9. The Head,
   Analytical Chemistry Section
   Applied Toxicology Laboratory
   Istituto Superiore di Sanità
   Attn: Dr S Caroli
   Viale Regina Elena 299
   I-00161 ROMA
   Tel: (6)4990
   Fax: (6)4440112

Responsible organisations

10. Dr. Alfredo Di Domeni-Cantonio
Infoterra National Focal Point
Ministero dell'Ambiente Servizio Via
Via della Ferratella In Laterano, 33
I-00184 ROMA RM
Tel: (396) 7036 2309
Fax: (396) 7025 7005

11. Associazione Industrie Difesa Produzioni Agricole (AGROFARMA)
Via Accademia 33
I-20131 MILANO
Tel: (39 2) 2681 0229
Fax: (39 2) 2681 0331

Jamaica

Responsibility	
Main authority responsible for approval/registration of pesticides	1
Main authority responsible for pesticide use directives, permits and training	1, 11
Main authority responsible for establishing national maximum pesticide residue limits	1, 8
Main authority responsible for control of pesticide residues in food	1, 8
Main laboratory for analysis of pesticide residues in food	8
Main laboratory for quality control of pesticide products	3, 6
Main organisation responsible for applied research regarding pesticide use. Information on related regional research	1, 2, 4, 5
Designated national authority for notifications and information regarding banned and severely restricted pesticides including the PIC procedure	1
Codex Alimentarius contact point	1
National IRPTC correspondent	1
Infoterra national focal point	7
National IFCS focal point	3
National association of pesticide manufacturers	9, 10

Responsible organisations

1. The Registrar
   Pesticide Control Authority
   Ministry of Health
   P.O. Box 472
   KINGSTON 5
   Tel: (809) 926 9220
   Fax: (809) 926 3857

2. Plant Protection Division
   Ministry of Agriculture
   Bodles Research Station
   Old Harbour P.O.
   ST. CATHERINE
   Tel: (809) 983 2267, 983 2242
   Fax: (809) 983 2281, 983 2243

3. Goverment Chemist
   Hope Gardens
   KINGSTON 6
   Tel: 927 1829

4. University of the West Indies
   Department of Entomology
   Mona
   KINGSTON 7
   Tel: 927 1202

5. Caribbean Agriculture Research and Development Institute
   University of the West Indies
   Mona
   KINGSTON 7
   Tel: 927 1231
   Fax: 977 1222

6. Bureau of Standards
   P.O. Box 113
   KINGSTON 10
   Tel: (809) 926 3140
   Fax: (809) 929 4736

7. Ms. Yolanda N. Mitto
   Infoterra NFP
   Natural Resources Conservation Authority
   P.O. Box 305
   KINGSTON 10
   Tel: (809) 923 5166
   Fax: (809) 923 5070

8. Food Storage and Prevention of Infestation Division
   Ministry of Industry and Commerce
   15 Gordon Town Road
   KINGSTON 6
   Tel: 927 1929, 927 1930

9. Jamaica Agricultural Chemical Association
   c/o 2 Dilcoosha Avenue
   KINGSTON 10
   Tel: 968 4702

10. Jamaica Manufacturers Association Ltd.
    Chemical Section
    85a Duke Street
    KINGSTON
    Tel: 922 8880, 922 2365

11. Rural Agricultural Development Authority
    Ministry of Agriculture
    Hope Gardens
    KINGSTON 6
    Tel: 977 1158

Japan
Nihon

Responsibility

Main authority responsible for approval/registration of pesticides	1, 2
Main authority responsible for pesticide use directives, permits and training	
Main authority responsible for establishing national maximum pesticide residue limits	3, 4
Main authority responsible for control of pesticide residues in food	3, 5
Main laboratory for analysis of pesticide residues in food	5, 6
Main laboratory for quality control of pesticide products	
Main organisation responsible for applied research regarding pesticide use. Information on related regional research	6, 7
Designated national authority for notifications and information regarding banned and severely restricted pesticides including the PIC procedure	8, 9
Codex Alimentarius contact point	10
National IRPTC correspondent	11
Infoterra national focal point	12
National IFCS focal point	14
National association of pesticide manufacturers	13

Note

Poisons information/control centre: 1-4-2 Shinsenri, Higashi-machi, Senri Life Sciences Center Building 11F, Toyonaka City, Osaka 565, Tel: (81 6) 871 8999, Fax: (81 6) 873 2041 [Head: Dr. S. Shintani]

Responsible organisations

1. Plant Protection Division
   Ministry of Agriculture,
   Forestry and Fisheries
   1-2-1 Kasumigaseki
   Chiyoda-ku
   TOKYO 100
   Tel: (3)5028111

2. Agricultural Chemicals
   Inspection Station
   Ministry of Agriculture,
   Forestry and Fisheries
   2-772 Suzuki-cho
   Kodaira-shi
   TOKYO 187
   Tel: (423)832151

3. Ministry of Health and
   Welfare
   1-2-2- Kasumigaseki,
   Chiyoda-ku
   TOKYO 100
   Tel: (3)5031711

4. Soil and Agricultural
   Chemicals Division
   Environment Agency
   1-2-2 Kasumigaseki
   Chiyoda-ku
   TOKYO 100
   Tel: (3)5813351

5. National Institute of Hygienic
   Sciences
   1-18-1 Kamiyoga
   Setagaya-ku
   TOKYO 158
   Tel: (3) 7001141

6. Institute of Environmental
   Toxicology
   2-772 Suzuki-cho
   Kodaira-shi
   TOKYO 187
   Tel: (423)837641

7. National Institute of Agro-
   Environmental Sciences
   3-3-1 Kannon-dai
   Yatabe
   TSUKUBA-IBARAKI 305
   Tel: (2975)68148

8. The Director
   Plant Protection Division
   Ministry of Agriculture,
   Forestry and Fisheries
   Kasumigaseki 1-2-1
   Chiyoda-ku
   TOKYO

9. The Director
   Global Issues Division
   Multilateral Cooperation
   Department
   Ministry of Foreign Affairs
   2-2-1 Kasumigaseki
   Chiyoda-ku
   TOKYO 100
   Tel: (81 3) 581 3882
   Fax: (81 3) 592 0364

10. Mr. K Tsukamoto
    Director, Resources Office
    Science and Technology Policy
    Bureau
    Science and Technology
    Agency
    2-2-1 Kasumigaseki
    Chiyoda-ku
    TOKYO 100
    Tel: (3)35815271
    Fax: (3)35813079

11. The Director General
National Institute of Hygiene Sciences
18-1 Kamiyoga, 1-chome
Setagaya-ku
TOKYO 158

12. Mr. Hisoshi Hatano
Infoterra NFP
Director,
Environmental Information Centre
National Institute for Environmental Studies
16-2 Onogawa
TSUKUBA
IBARAKI 305
Tel: (81 298) 516 111
Fax: (81 298) 514 732

13. Society of Agricultural Chemical Industry (SACI)
Nihonbashi Club Building
1-8 Nihonbashi Muromachi
Chuo-ku
TOKYO 103
Tel: (81 3) 3241 0215,
3241 0230
Fax: (81 3) 3241 3149

14. Mr K. Uchida
Director,
Office of Environmental Chemical Safety
Ministry of Health and Welfare
1-2-2 Kasumigaseki
Chiyoda-ku
TOKYO 100-45
Tel: (81 3) 3593 8913
Fax: (81 3) 3595 2298

Jordan

Al-Mamlakah al-Urdunniyah al-Hashimiyah

Responsibility	
Main authority responsible for approval/registration of pesticides	1, 2, 3, 4
Main authority responsible for pesticide use directives, permits and training	
Main authority responsible for establishing national maximum pesticide residue limits	1
Main authority responsible for control of pesticide residues in food	2, 4, 5, 6
Main laboratory for analysis of pesticide residues in food	7
Main laboratory for quality control of pesticide products	
Main organisation responsible for applied research regarding pesticide use. Information on related regional research	5, 8
Designated national authority for notifications and information regarding banned and severely restricted pesticides including the PIC procedure	9
Codex Alimentarius contact point	10
National IRPTC correspondent	11
Infoterra national focal point	12
National IFCS focal point	-
National association of pesticide manufacturers	-

Responsible organisations

1. Ministry of Agriculture
   P.O. Box 2099
   AMMAN
   Tel: 639391

2. Ministry of Municipalities and Environment
   AMMAN

3. Ministry of Supply
   P.O. Box 830
   AMMAN
   Tel: 602121

4. Ministry of Health
   P.O. Box 86
   AMMAN
   Tel: 665131

5. Royal Scientific Society
   P.O. Box 925819
   AMMAN
   Tel:844701
   Fax: 962 6 844806

6. Jordan University
   Faculty of Agriculture
   Jordan University Road
   AMMAN
   Tel: 843555

7. Al Sumia Toxicology Laboratory in Baq a
   Ministry of Agriculture
   P.O. Box 2099
   AMMAN

8. Scientific Research Department
   Ministry of Agriculture
   P.O. Box 2099
   AMMAN

9. The Director
   Department of Plant Protection
   Ministry of Agriculture
   AMMAN
   Tel: (962 6) 686 151
   Fax: (962 6) 686 310

10. National Committee for Codex Alimentarius
    Directorate of Standards
    Ministry of Industry and Trade
    P.O. Box 2019
    AMMAN

11. The Director
    Environmental Research Centre
    Royal Scientific Society
    P.O. Box 925819
    AMMAN

12. Mr. Mahmoud Al-Omari
    Infoterra NFP
    Department of Environment
    P.O. Box 35206
    AMMAN
    Tel: (962 6) 672 131
    Fax: (962 6) 672 135

Kazakstan
Qazaqstan Respublikasi

Responsibility	
Main authority responsible for approval/registration of pesticides	1
Main authority responsible for pesticide use directives, permits and training	2
Main authority responsible for establishing national maximum pesticide residue limits	-
Main authority responsible for control of pesticide residues in food	3
Main laboratory for analysis of pesticide residues in food	4
Main laboratory for quality control of pesticide products	5
Main organisation responsible for applied research regarding pesticide use. Information on related regional research	5
Designated national authority for notifications and information regarding banned and severely restricted pesticides including the PIC procedure	6
Codex Alimentarius contact point	-
National IRPTC correspondent	7
Infoterra national focal point	8
National IFCS focal point	-
National association of pesticide manufacturers	-

Responsible organisations

1. GOSKHIMKOMISSIA -
The State Interdepartmental Commission on Testing and Registration of Chemical and Biological Means of Protection, Pheromones and Growth Regulators of Crops and Forest Plantations
59, Ryskulov Avenue
ALMATY 480026
Tel: (7 3272) 380 037

2. Republican Station of Crop Protection
59, Ryskulov Avenue
ALMATY 480026
Tel: (7 3272) 380 056

3. Ministry of Health
63, Ablai-khan Avenue
ALMATY 480003
Tel: (7 3272) 334 469
Fax: (7 3272) 331 719

4. Toxicology Laboratory
Republican Sanitary and Epidemiological Station
84, Auezov Avenue
ALMATY 480008
Tel: (7 3272) 530 055

5. Toxicology and Control Laboratory
Republican Station of Crop Protection
59, Ryskulov Avenue
ALMATY 480026
Tel: (7 3272) 380 124

6. Dr. G B Bektourova
Department of International Relations
Ministry of Ecology and Bioresources
106, Panfilov Street
ALMATY 480091
Tel: (7 3272) 631 760
Fax: (7 3272) 636 973

7. Ministry of Ecology and Bioresources
106, Panfilov Street
ALMATY 480091
Tel: (7 3272) 631 760
Fax: (7 3237) 636 973

8. Mr. B A Kembayev
Infoterra NFP
Kazak Institute of Scientific and Technical Information (KAZGOSINTI)
221, Bogenbai Batyr Street
ALMATY 480096
Tel: (7 3272) 421 204, 422 381
Fax: (7 3272) 428 059

Kenya

Jamhuri ya Kenya, Republic of Kenya

Responsibility	
Main authority responsible for approval/registration of pesticides	1
Main authority responsible for pesticide use directives, permits and training	
Main authority responsible for establishing national maximum pesticide residue limits	-
Main authority responsible for control of pesticide residues in food	2, 3, 5, 7, 8, 9, 10, 12, 13
Main laboratory for analysis of pesticide residues in food	5, 6, 7, 8, 9, 10, 11, 12, 13
Main laboratory for quality control of pesticide products	
Main organisation responsible for applied research regarding pesticide use. Information on related regional research	5, 6, 7, 8, 9, 10, 11, 12, 13
Designated national authority for notifications and information regarding banned and severely restricted pesticides including the PIC procedure	14
Codex Alimentarius contact point	15
National IRPTC correspondent	16
Infoterra national focal point	16
National IFCS focal point	4
National association of pesticide manufacturers	17

Note

Poisons information/control centre: University of Nairobi, College of Health Sciences, P.O. Box 57763, Nairobi, Tel: (254) 272 67 70 [Head: Dr J N Ombega]

Responsible organisations

1. Pest Control Products Board
P.O. Box 14733
NAIROBI
Tel: 48211, 49501

2. Kenya Bureau of Standards
P.O. Box 54974
NHC House
Harambee Avenue
NAIROBI
Tel: 541206

3. The Government Chemist
P.O. Box 30014
NAIROBI
Tel: 23302

4. Dr W D O Sakari
Director,
Occupational Health and Safety Services
Ministry of Labour and Manpower Development
Social Security House
P.O. Box 40326
NAIROBI
Tel: (254 2) 555 178
Fax: (254 2) 716 649

5. Katumani Research Station
P.O. Box 340
MACHAKOS
Tel: 21506

6. Cotton Research Station
Kibos
KISUMU
Tel: Kisumu 2938

7. Potato Research Station
Tigoni
P.O. Box 338
LIMURU
Tel: Karuri 40998

8. Maize Research Station
P.O. Box 450
KITALE
Tel: 20445

9. Horticultural Research Station
P.O. Box 220
THIKA
Tel: Thika 21281

10. National Wheat Research Station
NJORO
Tel: 26

11. Pyrethrum Research Station
P.O. Box 100
MOLO
Tel: Molo 41

12. Sugarcane Research Station
Kibos
P.O. Box 1221
KISUMU
Tel: Kisumu 2293

13. National Agricultural Laboratories
P.O. Box 14733
NAIROBI
Tel: 48211

14. The Secretary
Pest Control Products Board
Waiyaki Way
P.O. Box 14733
NAIROBI
Tel: (254 2) 443 369
Fax: (254 2) 44611

Responsible organisations

15. The Director
    Kenya Bureau of Standards
    P.O. Box 54974
    NHC House
    Harambee Avenue
    NAIROBI

16. Infoterra NFP
    National Environment
    Sccretariat (NES)
    Ministry of Environment and
    Natural Resources
    Kencom House
    P.O Box 67839
    NAIROBI
    Tel: 229261, 218079

17. Pesticide Chemicals
    Association of Kenya
    (PCAK)
    P.O. Box 60723
    NAIROBI
    Tel: (254 2) 230100
    Fax: (254 2) 227692

Kiribati

Republic of Kiribati

Responsibility

Main authority responsible for approval/registration of pesticides	
Main authority responsible for pesticide use directives, permits and training	
Main authority responsible for establishing national maximum pesticide residue limits	
Main authority responsible for control of pesticide residues in food	
Main laboratory for analysis of pesticide residues in food	
Main laboratory for quality control of pesticide products	
Main organisation responsible for applied research regarding pesticide use. Information on related regional research	
Designated national authority for notifications and information regarding banned and severely restricted pesticides including the PIC procedure	-
Codex Alimentarius contact point	-
National IRPTC correspondent	-
Infoterra national focal point	2
National IFCS focal point	1
National association of pesticide manufacturers	-

Responsible organisations

1. The Secretary for Foreign Affairs and International Trade
   Ministry of Foreign Affairs and International Trade
   P.O. Box 68
   Bairiki
   TARAWA
   Tel: (686) 21342
   Fax: (686) 21466

2. Infoterra NFP
   The Secretary
   Ministry of Environment and Social Development
   P.O. Box 64
   Bairiki, TARAWA
   Tel: (686) 21099
   Fax: (686) 21120

Korea, North
Choson Minjujuui In'min Konghwaguk, Democratic People's Republic of Korea

Responsibility	
Main authority responsible for approval/registration of pesticides	
Main authority responsible for pesticide use directives, permits and training	
Main authority responsible for establishing national maximum pesticide residue limits	
Main authority responsible for control of pesticide residues in food	
Main laboratory for analysis of pesticide residues in food	
Main laboratory for quality control of pesticide products	
Main organisation responsible for applied research regarding pesticide use. Information on related regional research	
Designated national authority for notifications and information regarding banned and severely restricted pesticides including the PIC procedure	-
Codex Alimentarius contact point	1
National IRPTC correspondent	-
Infoterra national focal point	2
National IFCS focal point	-
National association of pesticide manufacturers	-

Responsible organisations

1. Director
   Foodstuffs Institute
   P.O. Box 901
   PYONGYANG

2. Mr. Li Il Chul
   Infoterra NFP
   National Coordinating
   Committee for Environment
   P.O. Box 44
   Jungsong-dong
   Central District
   PYONGYANG
   Tel: (850 3) 338632
   Fax: (850 2) 814660

Korea, South
Taehan Minguk, Republic of Korea

Responsibility	
Main authority responsible for approval/registration of pesticides	1
Main authority responsible for pesticide use directives, permits and training	1
Main authority responsible for establishing national maximum pesticide residue limits	2
Main authority responsible for control of pesticide residues in food	2
Main laboratory for analysis of pesticide residues in food	6, 7
Main laboratory for quality control of pesticide products	4
Main organisation responsible for applied research regarding pesticide use. Information on related regional research	9
Designated national authority for notifications and information regarding banned and severely restricted pesticides including the PIC procedure	11, 12
Codex Alimentarius contact point	8
National IRPTC correspondent	3
Infoterra national focal point	5
National IFCS focal point	-
National association of pesticide manufacturers	10

Responsible organisations

1. The Director
   Sustainable Agriculture Division
   Ministry of Agriculture, Forestry and Fisheries
   Kyunggi-Do
   KWACHEON CITY
   Tel: (822) 503 7285
   Fax: (822) 503 7249

2. The Director
   Food Safety Division
   Ministry of Health and Social Affairs
   Kyunggi-Do
   KWACHEON CITY
   Tel: (822) 503 7586
   Fax: (822) 503 7534

3. The Director
   Toxic Substances Division
   Ministry of Environment
   Kyunggi-Do
   KWACHEON CITY
   Tel: (822) 504 9288
   Fax: (822) 504 9280

4. The Director
   Pesticide Quality Division
   Department of Crop Protection
   Agricultural Science and Technology Institute
   Rural Development Administration
   SUWON 441-707
   Tel: (82 331) 292 6215
   Fax: (82 331) 295 3448

5. Dr. Yoo-Won Lee
   Director,
   Infoterra NFP
   Research Planning Division
   National Institute of Environmental Research
   613-2, Bulkwang-Dong
   Eunpyong-Gu
   SEUOL
   Tel: (822) 389 6711
   Fax: (822) 389 8712

6. The Director
   Division of Food Chemistry
   National Institute of Health
   5 Nokbun-Dong, Eunpyung-Gu
   SEOUL 122-020
   Tel: (822) 380 1755
   Fax: (822) 382 4892

7. National Plant Quarantine Office
   480-2, Anyang-6 Dong
   Anyang City
   KYUNGGI-DO 430-016
   Tel: (82 343) 490 524

8. The Permanent Representative of the Republic of Korea to FAO
   Embassy of the Republic of Korea
   Via Barnaba Oriani 30
   I- 00197 ROME
   Italy
   Tel: (39 6) 808 8769

9. The Korean Society of Pathology
Plant Pathology Division
Department of Crop Protection
Agricultural Science and Technology Institute
Rural Development
Administration
SUWON 441-707
Tel: (82 331) 272 7322
Fax: (82 331) 266 9083

10. Korea Agricultural Chemical Industry Association (KACIA)
Taiguk Building
1358-9 Seocho-Dong
Seocho-Ku
SEOUL 137-070
Tel: (822) 567 1591
Fax: (822) 552 4134

11. The Director,
Imports Division
Ministry of Trade, Industries and Energy
1, Choongang-Dong
Kwachon City
KYUNGGI-DO 427-010
Tel: (822) 503 9440
Fax: (822) 502 1754

12. The Director,
Sustainable Agriculture, Division
Ministry of Agriculture, Forestry and Fisheries
1, Choonggang-Dong
Kwachon City,
KYUNGGI-DO 427-010
Tel: (822) 503 7284
Fax: (822) 503 3963

Kuwait
Dawlat al-Kuwayt

Responsibility	
Main authority responsible for approval/registration of pesticides	1
Main authority responsible for pesticide use directives, permits and training	4, 7
Main authority responsible for establishing national maximum pesticide residue limits	2, 3, 4
Main authority responsible for control of pesticide residues in food	3
Main laboratory for analysis of pesticide residues in food	3, 5
Main laboratory for quality control of pesticide products	3, 7
Main organisation responsible for applied research regarding pesticide use. Information on related regional research	2, 4, 7
Designated national authority for notifications and information regarding banned and severely restricted pesticides including the PIC procedure	2, 8
Codex Alimentarius contact point	2, 4
National IRPTC correspondent	2
Infoterra national focal point	2
National IFCS focal point	2
National association of pesticide manufacturers	-

Note

Poisons information/control centre: Amiri Hospital, P.O. Box 4077, Safat [Head: Dr. A. Assad]

Responsible organisations

1. Central Committee for Pesticide Control
   Ministry of Health
   P.O. Box 24395
   SAFAT 13104
   Fax: (965) 245 6836

2. Environment Protection Council
   P.O. Box 24395
   SAFAT 13104
   Tel: (965) 245 2790. 245 6835
   Fax: (965) 242 1993, 245 6836

3. Nutrition and Food Administration
   Mionistry of Health
   P.O. Box 4078
   SAFAT 13041
   Tel: (965) 481 6043, 484 1385
   Fax: (965) 483 6155

4. Agricultural Affairs and Fish Resources Authority
   P.O. Box 21422
   SAFAT 13075
   Tel: (965) 242 7161
   Fax: (965) 473 9148

5. Environment Protection Department
   Ministry of Health
   P.O. Box 24395
   SAFAT 13104
   Tel: (965) 245 6833
   Fax: (965) 245 6836

6. Kuwait Institute for Scientific Research
   P.O. Box 24885
   SAFAT 13109
   Tel: (965) 483 6100, 481 6988
   Fax: (965) 484 6891

7. Ministry of Health
   Preventive Health Directorate
   Pest Control Section
   P.O. Box 4083
   SAFAT
   Fax: (965) 246 7852

8. The Director
   Plant Wealth Department
   Public Authority for Agricultural Affairs and Fish Resources
   P.O. Box 21422
   SAFAT 13075
   Tel: (965) 242 7161

Kyrgyzstan
Kyrgyzstan Respublikasy

Responsibility	
Main authority responsible for approval/registration of pesticides	1
Main authority responsible for pesticide use directives, permits and training	2, 5
Main authority responsible for establishing national maximum pesticide residue limits	2, 4
Main authority responsible for control of pesticide residues in food	3, 4
Main laboratory for analysis of pesticide residues in food	3, 4
Main laboratory for quality control of pesticide products	-
Main organisation responsible for applied research regarding pesticide use. Information on related regional research	2, 4
Designated national authority for notifications and information regarding banned and severely restricted pesticides including the PIC procedure	-
Codex Alimentarius contact point	-
National IRPTC correspondent	-
Infoterra national focal point	5
National IFCS focal point	-
National association of pesticide manufacturers	-

Responsible organisations

1. National Committee on Environmental Protection
   Isanov Street 131
   BISHKEK 720033
   Tel: (3312) 219 737
   Fax: (3312) 214 845

2. Ministry of Agriculture
   Plant Protection Administration
   Umetaliev Street 34
   BISHEK 720010
   Tel: (3312) 247 008

3. Kyrgyzstandard - National Inspection of Standardization and Metrology
   Panfilov Street 197
   BISHKEK 720040
   Tel: (3312) 264 862
   Fax: (3312) 264 708

4. National Sanitary-epidemiological Station
   Frunze Street 535
   BISHKEK 720033
   Tel: (3312) 262 214
   Fax: (3312) 265 809

5. Mr. K Sydykov
   Infoterra NFP
   National Information Centre
   Frunze Street 421
   BISHKEK 720011
   Tel: (3312) 299 658, 299 626
   Fax: (3312) 299 622

Laos

Sathalanalat Paxathipatai Paxaxôn Lao

Responsibility	
Main authority responsible for approval/registration of pesticides	1
main authority responsible for pesticide use directives, permits and training	
Main authority responsible for establishing national maximum pesticide residue limits	1
Main authority responsible for control of pesticide residues in food	2
Main laboratory for analysis of pesticide residues in food	2
Main laboratory for quality control of pesticide products	
Main organisation responsible for applied research regarding pesticide use. Information on related regional research	1, 3
Designated national authority for notifications and information regarding banned and severely restricted pesticides including the PIC procedure	-
Codex Alimentarius contact point	-
National IRPTC correspondent	-
Infoterra national focal point	4
National IFCS focal point	-
National association of pesticide manufacturers	-

Responsible organisations

1. Science, Technology and Environments Organization
   The Prime Minister's Office
   P.O. Box 2279
   VIENTIANE
   Tel: 2436
   Fax: 2433

2. Commission on Food and Drug Administration
   Ministry of Public Health
   VIENTIANE

3. Plant Protection Service
   Ministry of Agriculture and Forestry
   VIENTIANE

4. Mr. Pho Muangnalad
   Infoterra NFP
   Director of Cabinet
   Science, Technology and Environment
   Prime Minister's Office
   P.O. Box 2279
   VIENTIANE
   Tel: (856 21) 3470
   Fax: (856 21) 3472

Latvia

Latvijas Republika

Responsibility	
Main authority responsible for approval/registration of pesticides	1
Main authority responsible for pesticide use directives, permits and training	1
Main authority responsible for establishing national maximum pesticide residue limits	1
Main authority responsible for control of pesticide residues in food	4, 6
Main laboratory for analysis of pesticide residues in food	10
Main laboratory for quality control of pesticide products	7
Main organisation responsible for applied research regarding pesticide use. Information on related regional research	8
Designated national authority for notifications and information regarding banned and severely restricted pesticides including the PIC procedure	5, 9
Codex Alimentarius contact point	4
National IRPTC correspondent	5
Infoterra national focal point	2
National IFCS focal point	3
National association of pesticide manufacturers	-

Note

Poisons information/control centre: Republican Toxicological Center, 7th Riga Hospital, Hippocrat Str 2, Riga, Tel: 371(2) 536660, Fax: 371(2) 539524 [Head. Dr Viesturs Liguts]

Responsible organisations

1. State Plant Protection Station
   2 Republikas lauk.
   LV-1981 RIGA
   Tel: (371 2) 732 5098
   Fax: (371 9) 346 598

2. Dr. Guntis Pukitis
   Latvian Environmental Data Center
   2 Straumes Str.
   LV-2015 JURMALA
   Tel: (371 2) 762 282, 764 439
   Fax: (371 2) 764 439, 228 159

3. Dr. A Roska
   Head, Chemicals Inspectorate
   Environmental State Inspection
   27 Rupniecibas Str.
   LV-1877 RIGA
   Tel: (371 2) 325 091, 325 195
   Fax: (371 8) 820 442

4. Ms.Eugenia Pushmucane
   Head, State Inspection of Product Quality
   Ministry of Agriculture
   2 Republikas lauk.
   LV-1981 RIGA
   Tel: (371 2) 323 833

5. Chief Toxicologist
   Latvian Toxicological Centre
   Riga Clinical Hospital No.7
   Hippocrat Str. 2
   LV-1038 RIGA
   Tel: (371 2) 53660
   Fax: (371 2) 439 524

6. Mr. Olafs Stengrevics
   Director of Health Department
   7 Klijanu Str.
   LV-1012 RIGA
   Tel: (371 2) 379 287
   Fax: (371 2) 379 231

7. VZRU "Raziba"
   14a Struktoru Str.
   LV-1039 RIGA
   Tel/Fax: (371 2) 552 996

8. Plant Protection Center
   36/38 Lielvardes Str.
   LV-1006 RIGA
   Tel: (371 2) 551 501

9. Deputy Director,
   State Plant Protection Station
   2, Republikas laukums
   LV-1981 RIGA
   Tel: (371 9) 346 598
   Fax: (371 9) 346 598

10. National Centre of Environmental Health
    7 Klijanu Street
    LV-1012 RIGA
    Tel: (371 2) 375 413

Lebanon
Al-Jumhuriyah al Lubnaniyah

Responsibility	
Main authority responsible for approval/registration of pesticides	1
Main authority responsible for pesticide use directives, permits and training	
Main authority responsible for establishing national maximum pesticide residue limits	
Main authority responsible for control of pesticide residues in food	
Main laboratory for analysis of pesticide residues in food	
Main laboratory for quality control of pesticide products	
Main organisation responsible for applied research regarding pesticide use. Information on related regional research	
Designated national authority for notifications and information regarding banned and severely restricted pesticides including the PIC procedure	1
Codex Alimentarius contact point	2
National IRPTC correspondent	-
Infoterra national focal point	3
National IFCS focal point	-
National association of pesticide manufacturers	

Responsible organisations

1. Monsieur le Directeur
   Comité national ICP
   Ministère de l'agriculture
   Building of the Green Project
   BEIRUT

2. LIBNOR - Institut Libanais des normes et spécifications
   B.P. 2806
   BEIRUT

3. Ms Amal Habib
   Infoterra NFP
   Director,
   Documentation Centre
   National Council for Scientific Research
   P.O. Box 11-8281
   BEIRUT
   Tel: (961 1) 822 636, 822 775
   Fax: (961 1) 822 639

Lesotho

Kingdom of Lesotho

Responsibility	
Main authority responsible for approval/registration of pesticides	
Main authority responsible for pesticide use directives, permits and training	
Main authority responsible for establishing national maximum pesticide residue limits	
Main authority responsible for control of pesticide residues in food	
Main laboratory for analysis of pesticide residues in food	
Main laboratory for quality control of pesticide products	
Main organisation responsible for applied research regarding pesticide use. Information on related regional research	
Designated national authority for notifications and information regarding banned and severely restricted pesticides including the PIC procedure	3
Codex Alimentarius contact point	1
National IRPTC correspondent	-
Infoterra national focal point	2
National IFCS focal point	-
National association of pesticide manufacturers	-

Responsible organisations

1. Mrs. M.N. Mpeta
   Director, Food and Nutrition
   Coordinating Office
   Private Bag A78
   MASERU 100

2. Infoterra NFP
   The Director,
   Institute of Education
   National University of Lesotho
   P.O. Roma 180
   MASERU
   Tel: (266) 340 601, 340 369
   Fax: (266) 340 000

3. National Environment Secretariat
   Prime Minister's Office
   P.O. Box 527
   MASERU
   Tel: (266) 316 604
   Fax: (266) 310 506, 310 102

Liberia
Republic of Liberia

Responsibility	
Main authority responsible for approval/registration of pesticides	1
Main authority responsible for pesticide use directives, permits and training	
Main authority responsible for establishing national maximum pesticide residue limits	-
Main authority responsible for control of pesticide residues in food	-
Main laboratory for analysis of pesticide residues in food	-
Main laboratory for quality control of pesticide products	
Main organisation responsible for applied research regarding pesticide use. Information on related regional research	-
Designated national authority for notifications and information regarding banned and severely restricted pesticides including the PIC procedure	1
Codex Alimentarius contact point	3
National IRPTC correspondent	1
Infoterra national focal point	4, 5
National IFCS focal point	-
National association of pesticide manufacturers	2

Responsible organisations

1. The Director
   Division of Environmental and Occupational Health
   Ministry of Health and Social Welfare
   P.O. Box 10-9009
   1000 MONROVIA 10
   Tel: (231) 222 633, 224 221

2. Ministry of Health and Social Welfare
   P.O. Box 10-9009
   1000 MONROVIA 10

3. Mr. J M Coleman
   Director of Standards
   Ministry of Commerce and Industry
   P.O. Box 10-9041
   1000 MONROVIA

4. Hon. Thomas N. Brima
   Infoterra NFP
   Minister,
   Ministry of Internal Affairs
   P.O. Box 9008
   MONROVIA

5. Hon. James H. Lewis, Jr.
   Acting Assistant Minister
   Infoterra NFP
   Research and Development Planning
   Ministry of Internal Affairs
   P.O. Box 9008
   MONROVIA

Libya

al-Jamahiriyah al-Arabiyah al-Libyah ash-Shabiyah al-Ishtirakiyah, Libyan Arab Jamahiriya

Responsibility	
Main authority responsible for approval/registration of pesticides	1
Main authority responsible for pesticide use directives, permits and training	1, 4
Main authority responsible for establishing national maximum pesticide residue limits	2
Main authority responsible for control of pesticide residues in food	1
Main laboratory for analysis of pesticide residues in food	1, 3
Main laboratory for quality control of pesticide products	1
Main organisation responsible for applied research regarding pesticide use. Information on related regional research	6
Designated national authority for notifications and information regarding banned and severely restricted pesticides including the PIC procedure	1, 5
Codex Alimentarius contact point	7
National IRPTC correspondent	1
Infoterra national focal point	8
National IFCS focal point	-
National association of pesticide manufacturers	-

Note

Poisons information/control centre: Centre for Health and Drug Research, P.O. Box 5111, Tripoli [Head: Dr. E. Ramadan]

Responsible organisations

1. Technical Centre for Environment Protection
Secretariat of Housing and Utility
P.O. Box 83618
TRIPOLI
Tel: (218 21) 444 8452, 444 5795, 444 6868
Fax: (218 21) 333 8098

2. National Centre for Standards and Metrology
P.O. Box 5178
309 80 TRIPOLI
Tel: 444 6937, 691 512

3. Industrial Research Centre
Food Laboratory
P.O. Box 3633
TRIPOLI
Tel: 691 512, 691 517

4. Technical Council for Agricultural Pest Control
P.O. Box 6445
71623 TRIPOLI
Fax: 38098

5 General Directorate of Plant Protection and Quarantine
Secretariat of Agriculture
P.O. Box 351
TRIPOLI
Tel: 333 9141

6. Agricultural Research Centre
P.O. Box 2480
TRIPOLI
Tel: 605 540
Fax: 605 541

7. Office of International Cooperation
Secretariat of Agricultural Reclamation and Land Development
c/o UNDP
P.O. Box 358
TRIPOLI

8. Dr Abdulfattah Boargob
Technical Centre for Environmenal Protection
P.O. Box 83618
TRIPOLI
Tel: (218 21) 444 5795, 444 8452
Fax: (218 21) 333 8098

Lithuania

Lietuvos Respublika

Responsibility	
Main authority responsible for approval/registration use of pesticides	1, 2, 3
Main authority responsible for pesticide use directives, permits and training	1, 2, 3
Main authority responsible for establishing national maximum pesticide residue limits	2, 4
Main authority responsible for control of pesticide residues in food	2, 4
Main laboratory for analysis of pesticide residues in food	2,4
Main laboratory for quality control of pesticide products	-
Main organisation responsible for applied research regarding pesticide use. Information on related regional research	5, 6, 8
Designated national authority for notifications and information regarding banned and severely restricted pesticides including the PIC procedure	9
Codex Alimentarius contact point	4
National IRPTC correspondent	7
Infoterra national focal point	11
National IFCS focal point	10
National association of pesticide manufacturers	-

Responsible organisations

1. Ministry of Environmental Protection
A. Juozapaviciaus 9
2600 VILNIUS
Tel: (370 2) 355 868, 358 275, 725 868, 728 275
Fax: (370 2) 358 020, 728 020

2. State Public Health Centre
Ministry of Health
Kalvariju Street 153
2042 VILNIUS
Tel: (370 2) 778 951
Fax: (370 2) 224 601, 765 281

3. State Plant Protection Station
Ministry of Agriculture
Ukmerges 41
2025 VILNIUS
Tel: (370 2) 625 737, 725 790
Fax: (370 2) 224 440, 725 791

4. Ass.Prof. J Maskeliunas
Director,
National Nutrition Centre
Kalvariju Street 153
2042 VILNIUS
Tel:(370 2) 778 919
Fax: (370 2) 778 913

5. Institute of Horticulture and Market-gardening
Babtai
4335 KAUNO RAJ.
Tel: (370 7) 206 8576

6. Institute of Agriculture
Dotnuva-Academy
5050 KEDAINIU RAJ.
Tel: (370 57) 37263, 37271
Fax: (370 57) 37333, 56996

7. Dr. Julius Ptashekas
Head,
Department of Monitoring and Programs
State Public Health Centre
Kalvariju Str. 153
2042 VILNIUS
Tel: (370 2) 779 041
Fax: (370 2) 765 281

8. Pesticide Control Laboratory
State Plant Protection Station
Savanoriu av. 287
3009 KAUNAS
Tel: (370 7) 771 737, 771 726
Fax: (370 7) 771 737

9. Chairman,
State Commission for Plant Protection Products Registration
Ministry of Agriculture
Gedimino Avenue 19
2025 VILNIUS
Tel: (370 2) 625 737
Fax: (370 2) 224 440

10. Dr. R Sabaliauskas
Director General,
State Public Health Centre
Kalvariju Street 153
2042 VILNIUS
Tel: (370 2) 778 951
Fax: (370 2) 765 281

11. Mr. Arturas Daubaras
Director General,
Joint Research Centre
Ministry of Environmental Protection
A. Juozapaviciaus 9
2600 VILNIUS
Tel: (370 2) 353 202, 723 202
Fax: (370 2) 358 020, 728 020

Luxembourg

Grand-Duché de Luxembourg, Groussherzogtum Letzebuerg, Grossherzogtum Luxemburg

Responsibility	
Main authority responsible for approval/registration of pesticides	1, 2, 3
Main authority responsible for pesticide use directives, permits and training	2
Main authority responsible for establishing national maximum pesticide residue limits	1, 4
Main authority responsible for control of pesticide residues in food	4
Main laboratory for analysis of pesticide residues in food	4
Main laboratory for quality control of pesticide products	6
Main organisation responsible for applied research regarding pesticide use. Information on related regional research	2
Designated national authority for notifications and information regarding banned and severely restricted pesticides including the PIC procedure	2
Codex Alimentarius contact point	4
National IRPTC correspondent	7
Infoterra national focal point	5
National IFCS focal point	5
National association of pesticide manufacturers	-

Responsible organisations

1. Ministère de la Santé
   57 Boulevard de la Pétrusse
   L-2320 LUXEMBOURG
   Tel: 478-1
   Fax: 484 903

2. Administration des services techniques de l'agriculture
   B.P. 1904
   L-1019 LUXEMBOURG
   Tel: 457 172-1
   Fax: 4571 72341

3. Ministère de l'Agriculture, de la Viticulture et du Dévelopment Rural
   1 rue de la Congrégation
   L-1352 LUXEMBOURG
   Tel: 478-1
   Fax: 464 027

4. Laboratoire National de Santé
   Division du Contrôle des Denrées Alimentaires
   B.P. 1102
   L-1011 LUXEMBOURG
   Tel: 491 191
   Fax: 494 258

5. Ministère de l'Environnement
   18, Montée de la Pétrusse
   L-2918 LUXEMBOURG
   Tel: 478 6824
   Fax: 400 410

6. Laboratoires de Controle et d'Essais
   B.P. 75
   L-9001 ETTELBRUCK
   Tel: 82109
   Fax: 816 286

7. Laboratoire National de la Santé
   Attn: Dr. R Wennig
   B.P. 1102
   L-1011 LUXEMBOURG
   Tel: 494 939
   Fax: 404 319

Macedonia
Republika Makedonija

Responsibility

Main authority responsible for approval/registration of pesticides	
Main authority responsible for pesticide use directives, permits and training	
Main authority responsible for establishing national maximum pesticide residue limits	
Main authority responsible for control of pesticide residues in food	
Main laboratory for analysis of pesticide residues in food	
Main laboratory for quality control of pesticide products	
Main organisation responsible for applied research regarding pesticide use. Information on related regional research	
Designated national authority for notifications and information regarding banned and severely restricted pesticides including the PIC procedure	-
Codex Alimentarius contact point	1
National IRPTC correspondent	-
Infoterra national focal point	-
National IFCS focal point	-
National association of pesticide manufacturers	-

Responsible organisations

1. National Institute for Health Protection
   5o Udarna Divizija Str. No. 6
   91000 SKOPJE

Madagascar

Repoblikan'i Madagasikara, République de Madagascar

Responsibility	
Main authority responsible for approval/registration of pesticides	1
Main authority responsible for pesticide use directives, permits and training	
Main authority responsible for establishing national maximum pesticide residue limits	1
Main authority responsible for control of pesticide residues in food	1, 2, 3
Main laboratory for analysis of pesticide residues in food	1, 2, 3
Main laboratory for quality control of pesticide products	
Main organisation responsible for applied research regarding pesticide use. Information on related regional research	4
Designated national authority for notifications and information regarding banned and severely restricted pesticides including the PIC procedure	5
Codex Alimentarius contact point	6
National IRPTC correspondent	7, 8
Infoterra national focal point	9
National IFCS focal point	10
National association of pesticide manufacturers	-

Note

Poisons information/control centre: Centre Anti-Poisons, Service de Reanimation Medicale et de Toxicologie Clinique, Hopital Ravoahangy Andrianavalona, Ampefiloha, Antananarivo 101, Tel: (261) 227979, [Head: Dr (Ms) A Ramialiharisoa]

Responsible organisations

1. Service de la Protection des Végétaux
Ministère de la Production Agricole et de la Réforme Agraire
Ambatobe
TANANARIVE 101
Tel: 40610, 40205

2. Laboratoire des Fraudes Alimentaires
Ministère de la Santé Publique
TANANARIVE
Tel: 22239

3. Centre National de Recherches
Ministère de la Recherche Scientifique
Androhibe
TANANARIVE
Tel: 42802

4. FOFIA
Division Pathologie Végétale
Ambatobe
TANANARIVE
Tel: 40278

5. Monsieur le Directeur
Office National de l'Environnement
Lot IG 161A Ambalavao Isotry
TANANARIVE 101
Tel: 25999
Fax: 30693

6. Direction de la Qualite et de la Metrologie Legal
Ministère du Commerce
B.P. 1316
101 TANANARIVE

7. Monsieur le Chef du Département de Chimie
Centre National de Recherches Pharmaceutiques (CNRP)
B.P. 702
TANANARIVE

8. Monsieur le Directeur
Laboratoire de Biochimie
Service de Biologie Végétale et de Bichimie
E.E.S. Sciences
Université de Madagascar
B.P. 906
TANANARIVE

9. Mr. Finoano Henri
PFN d'Infoterra
Direction des Eaux et Forêts
Nanisana
B.P. 243
TANANARIVE 101
Tel: (261 2) 40811
Fax: (261 2) 40230

10. Prof. G Refeno
Directeur,
Centre National de Recherche sur l'Environnement
TANANARIVE
Tel: (261-2) 23962
Fax: (261-2) 26469

Malawi
Republic of Malawi

Responsibility	
Main authority responsible for approval/registration of pesticides	1
Main authority responsible for pesticide use directives, permits and training	1, 5
Main authority responsible for establishing national maximum pesticide residue limits	1
Main authority responsible for control of pesticide residues in food	2
Main laboratory for analysis of pesticide residues in food	2
Main laboratory for quality control of pesticide products	-
Main organisation responsible for applied research regarding pesticide use. Information on related regional research	1
Designated national authority for notifications and information regarding banned and severely restricted pesticides including the PIC procedure	5
Codex Alimentarius contact point	2
National IRPTC correspondent	1, 3
Infoterra national focal point	3
National IFCS focal point	-
National association of pesticide manufacturers	4

Responsible organisations

1. The Officer-In-Charge
   Bvumbwe Agricultural
   Research Station
   P.O. Box 5748
   LIMBE
   Tel: (265) 662 206
   Fax: (265) 662 323

2. The General Manager
   Malawi Bureau of Standards
   P.O. Box 946
   BLANTYRE
   Tel: 670 488
   Fax: 670 756

3. Infoterra NFP
   Attn: Mrs. E J Mede
   Ministry of Research and
   Environmental Affairs
   P.O. Box 30745
   LILONGWE 3
   Tel: (265) 781 111
   Fax: (265) 781 487

4. Pesticide Suppliers Association
   of Malawi
   P.O. Box 30188
   LILONGWE 3
   Tel: (265) 765 240
   Fax: (265) 765 384

5. Ministry of Research and
   Environmental Affairs
   Office of the President and
   Cabinet
   P.O. Box 30745
   LILONGWE 3
   Tel: (265) 781 111
   Fax: (265) 781 487

Malaysia

Responsibility

Main authority responsible for approval/registration of pesticides	1
Main authority responsible for pesticide use directives, permits and training	1
Main authority responsible for establishing national maximum pesticide residue limits	1, 2
Main authority responsible for control of pesticide residues in food	2
Main laboratory for analysis of pesticide residues in food	1, 4
Main laboratory for quality control of pesticide products	1, 4
Main organisation responsible for applied research regarding pesticide use. Information on related regional research	1, 3, 5, 6
Designated national authority for notifications and information regarding banned and severely restricted pesticides including the PIC procedure	1
Codex Alimentarius contact point	9
National IRPTC correspondent	7
Infoterra national focal point	7
National IFCS focal point	7
National association of pesticide manufacturers	8

Note

Poisons information/control centre: National Poisons Centre, University Sains Malaysia, Minden, 11800 Penang, Tel: (604) 657 0099, 657 7888, Fax: (604) 656 8417

Responsible organisations

1. The Secretary,
   Pesticides Board
   Department of Agriculture
   Jalan Gallagher
   50480 KUALA LUMPUR
   Tel: (60 3) 298 3077
   Fax: (60 3) 298 3646

2. Director,
   Food Quality Control Unit
   (Public Health)
   Ministry of Health
   4th Floor, Block E
   Jalan Dungun
   Damansara Heights
   50490 KUALA LUMPUR
   Tel: (60 3) 254 0088
   Fax: (60 3) 253 7804

3. Director-General
   Malaysian Agricultural
   Research and Development
   Institute (MARDI)
   P.O. Box 123301
   Postal Headquarters
   50774 KUALA LUMPUR
   Tel: (60 3) 943 7111
   Fax: (60 3) 948 3664

4. Director General of Chemistry
   Department of Chemistry
   Jalan Sultan
   46661 PETALING JAYA
   Tel: (60 3) 756 9522
   Fax: (60 3) 755 6764

5. Director General,
   Palm Oil Research Institute of
   Malaysia (PORIM)
   6, Persiaran Institusi
   Bandar Baru Bangi
   4300 KAJANG
   Tel: (60 3) 825 9155
   Fax: (60 3) 825 9446

6. Director,
   Rubber Research Institute of
   Malaysia (RRIM)
   P.O. Box 10150
   260, Jalan Ampang
   50450 KUALA LUMPUR
   Tel: (60 3) 456 7033
   Fax: (60 3) 457 3512

7. Director General,
   Department of Environment
   13th Floor, Wisma Sime
   Darby
   Jalan Raja Laut
   50662 KUALA LUMPUR
   Tel: (60 3) 294 7844
   Fax: (60 3) 293 1480,
   293 6146

8. Malaysian Agricultural
   Chemicals Association
   (MACA)
   TST 43, Damansara Jaya
   47400 PETALING JAYA
   Tel: (60 3) 704 8968
   Fax: (60 3) 704 8964

9. Malaysian National Codex
   Committee
   Standards and Industrial
   Research Institute of
   Malaysia (SIRIM)
   P.O. Box 7035
   Persiaran Dato'Menteri
   Seksyen 2
   40911 SHAH ALAM
   Tel: (60 3) 559 2601
   Fax: (60 3) 550 8095

Maldives
Divehi Jumhuriyya

Responsibility

Main authority responsible for approval/registration use of pesticides	
Main authority responsible for pesticide use directives, permits and training	
Main authority responsible for establishing national maximum pesticide residue limits	
Main authority responsible for control of pesticide residues in food	
Main laboratory for analysis of pesticide residues in food	
Main laboratory for quality control of pesticide products	
Main organisation responsible for applied research regarding pesticide use. Information on related regional research	
Designated national authority for notifications and information regarding banned and severely restricted pesticides including the PIC procedure	-
Codex Alimentarius contact point	1
National IRPTC correspondent	1
Infoterra national focal point	2
National IFCS focal point	-
National association of pesticide manufacturers	

Responsible organisations

1. Dr Abdul Samad Abdullah
   Director of National Health Services
   Ministry of Health
   MALE

2. Mr. Hussain Shihab
   Infoterra NFP
   Deputy Minister,
   Ministry of Planning, Human Resources and Environment
   Ghaazee Building
   MALE
   Tel: (960 32) 4861
   Fax: (960 32) 7351

Mali

République du Mali

Responsibility	
Main authority responsible for approval/registration of pesticides	
Main authority responsible for pesticide use directives, permits and training	
Main authority responsible for establishing national maximum pesticide residue limits	
Main authority responsible for control of pesticide residues in food	
Main laboratory for analysis of pesticide residues in food	
Main laboratory for quality control of pesticide products	
Main organisation responsible for applied research regarding pesticide use. Information on related regional research	
Designated national authority for notifications and information regarding banned and severely restricted pesticides including the PIC procedure	-
Codex Alimentarius contact point	-
National IRPTC correspondent	3
Infoterra national focal point	1
National IFCS focal point	2
National association of pesticide manufacturers	

Responsible organisations

1. Mme. Nissama Haoua Coulibaly
PFN d'Infoterra
Chef de la Division Environnement
Direction nationale des eaux et forêts
Ministère de l'Environnement
B.P. 275
BAMAKO
Tel: (223) 225850, 225973

2. Dr. G Kanoute
Conseiller Technique
Ministère de la Santé, de la Solidarité et des Personnes agées
B.P. 232
BAMAKO
Tel: (223) 22 5301
Fax: (223) 224244

3. Monsieur le Directeur
Laboratoire Central Veterinaire
Ministere de l'environnement
BAMAKO

Malta

Responsibility	
Main authority responsible for approval/registration of pesticides	1
Main authority responsible for pesticide use directives, permits and training	10
Main authority responsible for establishing national maximum pesticide residue limits	2
Main authority responsible for control of pesticide residues in food	2, 3, 4
Main laboratory for analysis of pesticide residues in food	3, 4
Main laboratory for quality control of pesticide products	4
Main organisation responsible for applied research regarding pesticide use. Information on related regional research	1
Designated national authority for notifications and information regarding banned and severely restricted pesticides including the PIC procedure	5, 10
Codex Alimentarius contact point	7
National IRPTC correspondent	8
Infoterra national focal point	9
National IFCS focal point	6
National association of pesticide manufacturers	-

Responsible organisations

1. Ministry for Food, Agriculture and Fisheries
Barriera Wharf
VALLETTA CMR 02
Tel: 225236
Fax: 231294

2. Ministry for Social Policy
Merchants Street
VALLETTA CMR 02
Tel: 224071
Fax: 241289

3. Health Laboratory
Toxicology Laboratory Medical School
G'MANGIA
Tel: 221019
Fax: 235638

4. Standards Laboratory
Evans Building
Merchants Street
VALLETTA CMR 02
Tel: 221335, 221873
Fax: 236237

5. The Chief Government Medical Officer
Officer i/c of the Industrial Hygiene Unit
Department of Health
VALLETTA
Fax: (356) 234 494

6. Dr. Mark Gauchi
c/o Office of the Parliamentary Secretary for Human Resources
FLORIANA
Tel: (356) 331 286
Fax: (356) 344 767

7. Mr. J Samut
Standards Laboratory
Department of Industry
Evans Building
Merchants Street
VALLETTA CMR 02
Tel: (356) 221335
Fax: (356) 236237

8. Mr M Gauci
The Occupational Hygiene Officer
Department of Health and Environment
Industry Hygiene Unit
37/39 Rue D'Argens
MSIDA MSD 05
Tel: 331286
Fax: 344767

9. Mr. J. Grech
Infoterra NFP
Environment Protection Department
Education Building
FLORIANA CMR 02
Tel: (356) 231895
Fax: (356) 241378

10. Director of Agriculture
Research and Development Centre
Ghammieri
MARSA
Tel: (356) 492 138
Fax: (356) 440 251

Mauritania

al-Jumhuriyah al-Islamiyah al-Muritaniyah

Responsibility	
Main authority responsible for approval/registration of pesticides	
Main authority responsible for pesticide use directives, permits and training	
Main authority responsible for establishing national maximum pesticide residue limits	
Main authority responsible for control of pesticide residues in food	
Main laboratory for analysis of pesticide residues in food	
Main laboratory for quality control of pesticide products	
Main organisation responsible for applied research regarding pesticide use. Information on related regional research	
Designated national authority for notifications and information regarding banned and severely restricted pesticides including the PIC procedure	1
Codex Alimentarius contact point	-
National IRPTC correspondent	-
Infoterra national focal point	-
National IFCS focal point	-
National association of pesticide manufacturers	

Responsible organisations

1. Direction du Développement des Resources Agro-Pastorales
   Ministère du Développement Rural et de l'Environnement
   B.P.180
   NOUAKCHOTT
   Tel: (222 2) 57989, 57879
   Fax: (222 2) 58605

Mauritius

Republic of Mauritius

Responsibility	
Main authority responsible for approval/registration of pesticides	1
Main authority responsible for pesticide use directives, permits and training	1
Main authority responsible for establishing national maximum pesticide residue limits	2
Main authority responsible for control of pesticide residues in food	3
Main laboratory for analysis of pesticide residues in food	3, 4
Main laboratory for quality control of pesticide products	3
Main organisation responsible for applied research regarding pesticide use. Information on related regional research	5, 6
Designated national authority for notifications and information regarding banned and severely restricted pesticides including the PIC procedure	7
Codex Alimentarius contact point	8
National IRPTC correspondent	7, 9
Infoterra national focal point	10
National IFCS focal point	11
National association of pesticide manufacturers	-

Responsible organisations

1. Pesticide Control Board
Ministry of Health
PORT LOUIS
Tel: (230) 212 2110
Fax: (230) 208 0376

2. Ministry of Healt
PORT LOUIS

3. Chief Government Analyst
Ministry of Health
PORT LOUIS

4. Agricultural Chemistry Division
Ministry of Agriculture, Fisheries, Natural Resources and the Environment
REDUIT

5. Mauritius Sugar Industry Research Institute
REDUIT

6. Plant Pathology Division and Entomology Division
Ministry of Agriculture, Fisheries, Natural Resources and the Environment
REDUIT

7. The Chairman
Pesticide Control Board
Ministry of Health
Atchia Building
Sufren Street
PORT LOUIS
Tel: (230) 211 2847
Fax: (230) 208 0376

8. The Chief Agricultural Officer
Agricultural Services
Ministry of Agriculture, Fisheries and Natural Resources
NFP Building
Maillard Street
PORT LOUIS

9. The Director
Agricultural Chemistry Division
Ministry of Agriculture, Fisheries, Natural Resources and the Environment
REDUIT

10. Mr. R.H. Prayag
Director,
Department of Environment
Ministry of the Environment and Land Use
Ken Lee tower,
Line Barracks Street
PORT LOUIS
Tel: (230) 212 6080
Fax: (230) 212 6671

11. Dr R Sibartie
Ministry of Health
PORT LOUIS
Tel: (230) 211 2847
Fax: (230) 208 0376

Mexico
Estados Unidos Mexicanos

Responsibility

Main authority responsible for approval/registration of pesticides	1, 2, 3, 4
Main authority responsible for pesticide use directives, permits and training	1, 2, 3, 4
Main authority responsible for establishing national maximum pesticide residue limits	1, 2
Main authority responsible for control of pesticide residues in food	1, 2
Main laboratory for analysis of pesticide residues in food	1, 2
Main laboratory for quality control of pesticide products	-
Main organisation responsible for applied research regarding pesticide use. Information on related regional research	1
Designated national authority for notifications and information regarding banned and severely restricted pesticides including the PIC procedure	1, 2
Codex Alimentarius contact point	4
National IRPTC correspondent	1, 2
Infoterra national focal point	3
National IFCS focal point	2, 3
National association of pesticide manufacturers	5, 6

Note

Poisons information/control centres in Chihuahua and Monterrey

Responsible organisations

1. Direccion General de Sanidad Vegetal
Secretaría de Agricultura, Ganaderia y Desarollo Rural
Guillermo Pérez Valenzuela No. 127, 2 Piso
Col. El Carmen Coyoacan
C.P. 04100 MEXICO, D.F.
Tel: (52 5) 540 512,
658 1671
Fax: (52 5) 554 0529

2. Direccion General de Salud
Ambiental, Occupacional y Saneamiento Basico
Secretaria de Salud
San Luis Potosi No. 192,
4 Piso
Col. Roma
C.P. 06700 MEXICO, D.F.
Tel: (52 5) 584 6745
Fax: (52 5) 584 5260

3. Instituto Nacional de Ecologia
de la Secretaria de Medio Ambiente,
Recursos Naturales y Pesca
Rio Elba No.20, 12 Piso
Col. Cuauhtemoc
C.P. 06700 MEXICO, D.F.
Tel: (52 5) 286 6625
Fax: (52 5) 286 6625

4. Direccion General de Normas de la Secretariat de Comercio y Fomento Industrial
Puente de Tecamachalco No. 6
Lomas de Tecamachalco,
Seccion de Las Fuentes,
Naucalpan de Juarez
C.P. 53950 MEXICO, D.F.
Tel: (52 5) 540 2612, 520 8530
Fax: (52 5) 729 9484

5. Asociación Méxicana de la Industria de Plaguicidas y Fertilizantes (AMIPFAC)
Apartado Postal 60-594
C.P.11870 MEXICO, D.F.
Tel: (52 5) 598 9095
Fax: (52 5) 598 6666

6. Union Mexicana de Fabricantes y Formuladores de Agroquimicos, S.C. (UMFFASC)
Gabriel Mancera No. 1433, Depto.C-3
Esquina Felix Cuevas
Col. del Valle
C.P.03100 MEXICO, D.F.
Tel: (52 5) 601 1100
Fax: (52 5) 601 2619

Micronesia
Federated States of Micronesia

Responsibility	
Main authority responsible for approval/registration of pesticides	
Main authority responsible for pesticide use directives, permits and training	
Main authority responsible for establishing national maximum pesticide residue limits	
Main authority responsible for control of pesticide residues in food	
Main laboratory for analysis of pesticide residues in food	
Main laboratory for quality control of pesticide products	
Main organisation responsible for applied research regarding pesticide use. Information on related regional research	
Designated national authority for notifications and information regarding banned and severely restricted pesticides including the PIC procedure	-
Codex Alimentarius contact point	1
National IRPTC correspondent	-
Infoterra national focal point	-
National IFCS focal point	-
National association of pesticide manufacturers	

Responsible organisations

1. Dr. Eliuel K. Pretrick
   Secretary,
   Department of Health Services
   P.O. Box PS70
   PALIKIR, Pohnpei

Moldova

Rḗpublica Moldova, Republic of Moldova

Responsibility	
Main authority responsible for approval/registration of pesticides	3, 7
Main authority responsible for pesticide use directives, permits and training	1, 4, 5
Main authority responsible for establishing national maximum pesticide residue limits	1, 5
Main authority responsible for control of pesticide residues in food	4, 6
Main laboratory for analysis of pesticide residues in food	6
Main laboratory for quality control of pesticide products	9
Main organisation responsible for applied research regarding pesticide use. Information on related regional research	8
Designated national authority for notifications and information regarding banned and severely restricted pesticides including the PIC procedure	10
Codex Alimentarius contact point	-
National IRPTC correspondent	1
Infoterra national focal point	2
National IFCS focal point	-
National association of pesticide manufacturers	-

Responsible organisations

1. The Director,
   Principal Department of Hygiene and Epidemiology
   Ministry of Health
   St. Hincesti, 1
   277028 CHISINAU
   Tel: (373 2) 735 333, 729 716
   Fax: (373 2) 738 781

2. The Director,
   Ecological Monitoring Centre of the National Institute of Ecology
   State Department for Environment Protection and Natural Resources
   Bd. Dacia, 58
   277060 CHISINAU
   Tel: (373 2) 561 964, 570 488
   Fax: (373 2) 579 663

3. The Director,
   State Centre for Certification of Chemical and Biological Means of Plant Protection and Growth Regulators
   Ministry of Agriculture and Nourishment
   Bd. Stefan cel Mare, 162
   277012 CHISINAU
   Tel: (373 2) 248 490
   Fax: (373 2) 248 724

4. Ministry of Agriculture and Nourishment
   Bd. Stefan cel Mare, 162
   277012 CHISINAU
   Tel: (373 2) 233 427
   Fax: (373 2) 232 368

5. The Director General,
   State Department for Environmental Protection
   Bd. Stefan cel Mare, 73
   277001 CHISINAU
   Tel: (373 2) 233 806
   Fax: (373 2) 226 161

6. The Director General
   National Scientific Centre for Hygiene and Epidemiology
   Ministry of Health
   George Asachi, 67a
   277028 CHISINAU
   Tel: (373 2) 729 647
   Fax: (373 2) 729 725

7. The Director,
   State Ecological Inspectorate
   State Department for Environmental Protection
   Bd. Stefan cel Mare, 73
   277001 CHISINAU
   Tel: (373 2) 224 249
   Fax: (373 2) 265 440

8. The Director,
   Research and Technology Institute for Agrochemical Service
   Ministry of Agriculture and Nourishment
   Cosmonautilor, 6
   277005 CHISINAU
   Tel: (373 2) 243 066,
   243 472
   Fax: (373 2) 242 787

Responsible organisations

9. The Director,
   The Plant Protection Station
   Ministry of Agriculture and
   Nourishment
   Cartus Street, 77
   Durlesti
   278270 CHISINAU
   Tel: 723 963, 723 972
   Fax: (373 2) 232 368

10. The Chairman,
    Scientific Research Association
    "FERTILITATEA"
    6, Kosmonavtov Street
    277005 KISHINEV
    Tel: (373 2) 243 032
    Fax: (373 2) 242 787

Monaco
Principauté de Monaco

Responsibility

Main authority responsible for approval/registration of pesticides	
Main authority responsible for pesticide use directives, permits and training	
Main authority responsible for establishing national maximum pesticide residue limits	
Main authority responsible for control of pesticide residues in food	
Main laboratory for analysis of pesticide residues in food	
Main laboratory for quality control of pesticide products	
Main organisation responsible for applied research regarding pesticide use. Information on related regional research	
Designated national authority for notifications and information regarding banned and severely restricted pesticides including the PIC procedure	-
Codex Alimentarius contact point	-
National IRPTC correspondent	-
Infoterra national focal point	1
National IFCS focal point	-
National association of pesticide manufacturers	

Responsible organisations

1. Mr. Patrick van Klaveren
   Chef du Service de
   l'Environnement
   Departement des Travaux
   Publics et les Affaires
   Sociales
   3, Avenue de Fontvieille
   MC-98000 MONACO Cedex
   Tel: (33) 93158149, 93158963
   Fax: (33) 92052891

Mongolia
Mongol Uls

Responsibility	
Main authority responsible for approval/registration of pesticides	1, 5
Main authority responsible for pesticide use directives, permits and training	
Main authority responsible for establishing national maximum pesticide residue limits	1, 2
Main authority responsible for control of pesticide residues in food	2
Main laboratory for analysis of pesticide residues in food	2
Main laboratory for quality control of pesticide products	
Main organisation responsible for applied research regarding pesticide use. Information on related regional research	1, 2
Designated national authority for notifications and information regarding banned and severely restricted pesticides including the PIC procedure	6
Codex Alimentarius contact point	2
National IRPTC correspondent	3
Infoterra national focal point	4
National IFCS focal point	-
National association of pesticide manufacturers	-

Responsible organisations

1. Dr. Dambyn Tsedev
   Chief, Plant Protection Service
   Baruun Selve 9
   ULAANBAATAR 38
   Tel: 328684

2. Dr. Nagnain Saijaa
   State Inspection Department
   of Hygiene and
   Epidemiology
   Pease Street 49
   ULAANBAATAR 49
   Tel: 52677

3. The Chief,
   Environmental Monitoring
   Laboratory
   Ministry for Nature and
   Environment Protection
   Baigalyam
   Khudaldaany gudamg 5
   ULAANBAATAR 11

4. Dr. Mendbayaryn Badarch
   Infoterra National Focal Point
   Ministry for Nature and
   Environment
   Khudaldaany gudamg 5
   ULAANBAATAR 46/1144
   Tel: (976 1) 324198
   Fax: (976 1) 321401

5. Dr. Lodoisambuugyn
   Dolgormaa
   Expert, Ecological Expertise
   Division
   Ministry for Nature and
   Environment
   Khudaldaany gudamg 5
   ULAANBAATAR 11

6. Chief,
   State Plant Protection Service
   Ministry of Food and
   Agriculture
   P.O. Box 147
   ULAANBAATAR 38
   Tel: (976 1) 21670
   Fax: (976 1) 21670

Morocco
al-Mamlakah al-Maghribiyah

Responsibility

Main authority responsible for approval/registration of pesticides	1
Main authority responsible for pesticide use directives, permits and training	
Main authority responsible for establishing national maximum pesticide residue limits	1
Main authority responsible for control of pesticide residues in food	1
Main laboratory for analysis of pesticide residues in food	-
Main laboratory for quality control of pesticide products	
Main organisation responsible for applied research regarding pesticide use. Information on related regional research	1, 2, 3, 4
Designated national authority for notifications and information regarding banned and severely restricted pesticides including the PIC procedure	5
Codex Alimentarius contact point	6
National IRPTC correspondent	-
Infoterra national focal point	7
National IFCS focal point	-
National association of pesticide manufacturers	-

Note

Poisons information/control centre: Centre Anti-Poisons du Maroc, Institut National d'Hygiene, 27 avenue Ibn Batouta, BP 769, Rabat 11400, Tel: (212) 771902, Fax: (212) 731997 [Head: Ms Pr R Soulaymani Bencheikh]

Responsible organisations

1. Service de la Protection des Végétaux
   Direction de la Recherche Agronomique
   Ministère de lÁgriculture et de la Réforme Agraire
   B.P. 415
   RABAT

2. Section Entomologie et Phytopathologie
   Station Recherche Forestière
   B.P. 763
   RABAT

3. Institut National de la Recherche Agronomique
   B.P. 415
   RABAT

4. Département de Phytiatrie
   Institut Agronomique et Véterinaire Hassan II
   B.P. 704
   Agdal
   RABAT

5. Monseieur le Secrétaire général
   Ministère de l'agriculture et de la mise en valeur agricole
   Direction de la protection des végétaux, des contrôles techniques et de la répression des fraudes
   B.P. 1308
   RABAT
   Tel: (212 7) 771 078
   Fax: (212 7) 771 078

6. Division de la Répression des Fraudes
   Ministère de l'agriculture et de la mise en valeur agricole
   25, Avenue les Alaouiyines
   RABAT

7. M. Ahmed Fassi-Fihri
   Point Focal Infoterra
   Centre National de Documentation
   Hadj Ahmed Cherkaoui
   Haut Agdal
   B.P. 826
   RABAT
   Tel: (212 7) 773 131
   Fax: (212 7) 773 134

Mozambique

República de Moçambique

Responsibility	
Main authority responsible for approval/registration of pesticides	1
Main authority responsible for pesticide use directives, permits and training	
Main authority responsible for establishing national maximum pesticide residue limits	2
Main authority responsible for control of pesticide residues in food	2, 3
Main laboratory for analysis of pesticide residues in food	
Main laboratory for quality control of pesticide products	
Main organisation responsible for applied research regarding pesticide use. Information on related regional research	1
Designated national authority for notifications and information regarding banned and severely restricted pesticides including the PIC procedure	4
Codex Alimentarius contact point	5
National IRPTC correspondent	-
Infoterra national focal point	6
National IFCS focal point	-
National association of pesticide manufacturers	-

Responsible organisations

1. Ministério da Agricultura
   Instituto Nacional de
   Investigacáo
   Agronomica (INIA)
   P.O. Box
   MAPUTO
   Tel: 30091

2. Ministério de Saúde
   P.O. Box 264
   MAPUTO
   Tel: 27131
   Telex: 6239

3. Ministério da Agricultura
   Instituto Nacional de
   Veterinária (INV)
   P.O. Box
   MAPUTO
   Tel: 733031

4. Departamento de Sanidad
   Vegetal
   Direccao Nacional de
   Agricultura
   Av. das Forcas Populares de
   Liberacao de Mocambique
   C.P. 3658
   MAPUTO
   Tel: (258 1) 460 097
   Fax: (258 1) 417 141

5. Mr. R M de Melo
   Ministério de Saúde
   Secáo de Higiene, de Aguas e
   Alimentos
   P.O. Box 264
   MAPUTO

6. Ms. Jacinta Carlos
   Infoterra NFP
   Documentation and
   Information Centre
   Ministry for Coordination of
   Environmental Affairs
   (MICOA)
   Av. Acordos de Lusaka 2115
   P.O. Box 2020
   MAPUTO
   Tel: (258 1) 465 843, 465 708,
   466 058
   Fax: (258 1) 465849

Myanmar
Pyidaungzu Myanmar Naingngandaw

Responsibility	
Main authority responsible for approval/registration of pesticides	3
Main authority responsible for pesticide use directives, permits and training	1, 4
Main authority responsible for establishing national maximum pesticide residue limits	4
Main authority responsible for control of pesticide residues in food	2, 4
Main laboratory for analysis of pesticide residues in food	5
Main laboratory for quality control of pesticide products	5
Main organisation responsible for applied research regarding pesticide use. Information on related regional research	4, 6
Designated national authority for notifications and information regarding banned and severely restricted pesticides including the PIC procedure	8
Codex Alimentarius contact point	2
National IRPTC correspondent	
Infoterra national focal point	
National IFCS focal point	7
National association of pesticide manufacturers	-

Responsible organisations

1. The Director General
Department of Health
36, Theinbyu Road
YANGON

2. The Director
Food and Drug
Administration Department
35, Mawkundaik Road
YANGON
Tel: (95 01) 77291

3. Pesticide Registration Board
Myanma Agriculture Service
Kanbe Road
Yankin
YANGON 11081
Tel: (95 01) 63397, 67039
Fax: (95 01) 67991

4. Plant Protection Division
Myanma Agriculture Service
Bayinnaung Road
West Gyogone
Insein
YANGON 11011
Tel: (95 01) 63397
Fax: (95 01) 67991

5. Pesticide Analytical
Laboratory
Plant Protection Division
Myanma Agriculture Service
Bayinnaung Road
West Gyogone
Insein
YANGON 11011
Tel: (95 01) 63401
Fax: (95 01) 67991

6. Occupational Health Unit
Department of Health
36, Theinbyu Road
YANGON
Tel: (95 01) 91092

7. Dr. Win Lwin Nyunt
Head,
Occupational Health Unit
Department of Health
36, Theinbyu Road
YANGON
Tel: (95 01) 91092

8. Department of Agricultural
Planning
Ministry of Agriculture
Seinlei May Yeiktha Lane
Kabar Aye Pagoda Road
Yankin Township
YANGON
Tel: (95 1) 65842
Fax: (95 1) 63984

Namibia

Republic of Namibia

Responsibility	
Main authority responsible for approval/registration pesticides	1
Mainauthority responsible for pesticide use directives, permits and training	1
Main authority responsible for establishing national maximum pesticide residue limits	1
Main authority responsible for control of pesticide residues in food	1, 2
Main laboratory for analysis of pesticide residues in food	3
Main laboratory for quality control of pesticide products	-
Main organisation responsible for applied research regarding pesticide use. Information on related regional research	-
Designated national authority for notifications and information regarding banned and severely restricted pesticides including the PIC procedure	-
Codex Alimentarius contact point	-
National IRPTC correspondent	-
Infoterra national focal point	4
National IFCS focal point	5
National association of pesticide manufacturers	-

Responsible organisations

1. Chief Agricultural Inspector
   Department of Agriculture
   Put Bay 13184
   WINDHOEK

2. Director of Veterinary Services
   Put Bay 12022
   WINDHOEK

3. Veterinary Laboratory
   Department of Agriculture
   Put Bay 13184
   WINDHOEK

4. Dr. C J Brown
   Head, Environmental Affairs
   Directorate of Environmental Affairs
   Ministry of Environment and Tourism
   Private Bag 13306
   WINDHOEK
   Tel: (264 61) 249 105, 63131
   Fax: (264 61) 63195

5. Ministry of Environment and Tourism
   Directorate of Environmental Affairs
   Private Bag 13306
   WINDHOEK
   Tel: (264) 6124 9015
   Fax: (264) 6124 0339

Nepal

Nepál Adhirájya

Responsibility	
Main authority responsible for approval/registration of pesticides	1
Main authority responsible for pesticide use directives, permits and training	
Main authority responsible for establishing national maximum pesticide residue limits	2
Main authority responsible for control of pesticide residues in food	2
Main laboratory for analysis of pesticide residues in food	2
Main laboratory for quality control of pesticide products	
Main organisation responsible for applied research regarding pesticide use. Information on related regional research	3
Designated national authority for notifications and information regarding banned and severely restricted pesticides including the PIC procedure	4
Codex Alimentarius contact point	5
National IRPTC correspondent	-
Infoterra national focal point	6
National IFCS focal point	-
National association of pesticide manufacturers	-

Responsible organisations

1. Agriculture Input Corporation
   Central Office
   P.O. Box 195
   Teku
   KATHMANDU
   Tel: 13497

2. Quality Control and
   Standarisation Division
   Central Food Research
   Laboratory
   Babar Mahal
   KATHMANDU
   Tel: 12781

3. Division of Plant Pathology
   Department of Agriculture
   Khumaltar
   LALITPUR
   Tel: 21197

4. Chief,
   Plant Protection and Industrial
   Entomology Division
   Department of Agriculture
   Development
   Ministry of Agriculture
   Harihar Bhawan
   LALITPUR

5. Chief Food Research Officer
   Central Food Research
   Laboratory
   KATHMANDU
   Tel: 214 824, 212 781

6. Infoterra NFP
   National Committee for Man
   and the Biosphere
   National Commission for
   UNESCO
   Ministry of Education, Culture
   and Social Welfare
   Kaisher Mahal
   G.P.O. Box 1071
   KATHMANDU
   Tel: (977 1) 418 782
   Fax: (977 1) 412 460

Netherlands
Koninkrijk der Nederlanden

Responsibility	
Main authority responsible for approval/registration of pesticides	1
Main authority responsible for pesticide use directives, permits and training	
Main authority responsible for establishing national maximum pesticide residue limits	3, 4
Main authority responsible for control of pesticide residues in food	5, 6
Main laboratory for analysis of pesticide residues in food	6, 7
Main laboratory for quality control of pesticide products	
Main organisation responsible for applied research regarding pesticide use. Information on related regional research	6, 8, 9, 10, 11, 12, 13, 14
Designated national authority for notifications and information regarding banned and severely restricted pesticides including the PIC procedure	2
Codex Alimentarius contact point	16
National IRPTC correspondent	17
Infoterra national focal point	18
National IFCS focal point	-
National association of pesticide manufacturers	15

Note

Poisons information/control centre: National Institute of Public Health and Environmental Protection, Postbus 1, 3120 BA Bilthoven, Tel: (31) 3074 2710, Fax: (31) 3054 1511 [Head: Dr. J. Meulenbelt]

Responsible organisations

1. College voor de Toelating van Bestrijdingmiddelen
Postbus 217
6700 AE WAGENINGEN
Tel: (8370) 75478
Fax: (8370) 23193

2. The Director,
Chemicals Division
Ministry of Housing, Physical Planning and Environment
Directorate for Chemicals, External Safety and Radiation Protection
P.O. Box 30945
2500 GX DEN HAAG
Tel: (31 70) 339 4744
Fax: (31 70) 339 1297

3. Ministerie van Welzijn,
Volkgezondheid en Cultuur
Postbus 439
2260 AK LEIDSCHENDAM
Tel: (70) 209260

4. Ministerie van Landbouw en Visserij
Postbus 20401
2500 EK DEN HAAG
Tel: (70) 793911

5. Hoofdinspectie van de Volksgezondheid belast met het Toezicht op Levensmiddelen en Keuring van Waaren
Postbus 439
2260 AK LEIDSCHENDAM
Tel: (70) 209260

6. Rijksinstituut voor
Volksgezondheid en Milieuhygiene
Postbus 1
3720 BA BILTHOVEN
Tel: (30) 749 111
Fax: (31) 307 429 71

7. Rijks Kwaliteitsinstituut voor Land en Tuinbouwprodukten (RIKILT)
Postbus 230
6700 AE WAGENINGEN
Tel: (8370) 19110

8. Proefstation voor Tuinbouw onder Glas
Zuidweg 38
2671 MN NAALDWIJK
Tel: (1740) 26541

9. Proefstation voor Fruitteelt
Brugstraat 51
4475 AN WILHELMINADORP
Tel: (1100) 16390

10. Plantenziektenkundige Dienst
Postbus 9102
6700 HC WAGENINGEN
Tel: (8370) 19001

11. Instituut voor Onderzoek van Bestrijdingsmiddelen (IOB)
Marijkeweg 22
6709 PG WAGENINGEN
Tel: (8370) 11821

12. Fytopathologisch
 Laboratorium
 Javalaan 20
 3742 CP BAARN
 Tel: (2154) 15654

13. Instituut voor
 Plantenziektenkundig
 Onderzoek
 Wageningen Centre of
 Agricultural Science
 Postbus 9060
 6700 GW WAGENINGEN
 Tel: (8370) 19151

14. Proefstation voor de
 Akkerbouw en de
 Groenteteelt in Vollegrond
 (PAGV)
 Postbus 430
 8200 AK LELYSTAD
 Tel: (3200) 22714

15. Nederlandse Stichting voor
 Fytofarmacie (NEFYTO)
 Hogeweg 16
 Postbus 80523
 NL-2508 DEN HAAG
 Tel: (31 70) 351 4851
 Fax: (31 70) 354 9766

16. Executive Officer for Codex
 Alimentarius
 Nutrition and Quality Affairs
 Ministry of Agriculture,
 Nature Management and
 Fisheries
 Room 9327
 Postbus 20401
 2500 EK DEN HAAG
 Tel: (31 703) 792 456
 Fax: (31 703) 477 552

17. IRPTC-IPCS Working Group
 Ministry of Housing, Physical
 Planning and the Environment
 Chemicals Division
 Attn: Dr G P Hekstra
 Postbus 30945
 2500 GX DEN HAAG

18. Netherlands Government
 Information
 P.O. Box 51
 2501 CB DEN HAAG
 Tel: (31 6) 8501
 Fax: (31 70) 311 5177

New Zealand

New Zealand, Aotearoa

Responsibility	
Main authority responsible for approval/registration of pesticides	1
Main authority responsible for pesticide use directives, permits and training	1
Main authority responsible for establishing national maximum pesticide residue limits	2
Main authority responsible for control of pesticide residues in food	2
Main laboratory for control of pesticide residues in food	4
Main laboratory for quality control of pesticide products	10
Main organisation responsible for applied research regarding pesticide use. Information on related regional research	6
Designated national authority for notifications and information regarding banned and severely restricted pesticides including the PIC procedure	7
Codex Alimentarius contact point	8
National IRPTC correspondent	9
Infoterra national focal point	3
National IFCS focal point	9
National association of pesticide manufacturers	5

Note

Poisons information/control centre: National Hazardous Chemicals and Poisons Information Centre, University of Otago Medical School, P.O. Box 913, Dunedin, Tel: (643) 4797248, Fax: (643) 4747000 [Head: Dr. W. Temple]

Responsible organisations

1. Pesticides Board
c/o Agricultural Compounds Unit
Ministry of Agriculture
P.O. Box 40-063
UPPER HUTT
Tel: (64 4) 528 6089
Fax: (64 4) 528 4675

2. Ministry of Health
P.O. Box 5013
WELLINGTON
Tel: (64 4) 496 2000
Fax: (64 4) 496 2340

3. Infoterra NFP
Librarian in Charge
Ministry for the Environment
P.O. Box 10-362
WELLINGTON
Tel: (64 4) 498 7435
Fax: (64 4) 471 0195

4. National Chemical Residue Laboratory
Wallaceville Agricultural Research Centre
P.O. Box 40-063
UPPER HUTT
Tel: (64 4) 528 0718
Fax: (64 4) 528 0493

5. Agricultural Chemicals and Animal Remedies Manufacturers Association of New Zealand (AGCARM)
P.O. Box 5069
WELLINGTON
Tel: (64 4) 499 4255, 851 962
Fax: (64 4) 499 4223, 828 200

6. Ministry of Agriculture
P.O. Box 2526
WELLINGTON
Tel: (64 4) 474 4100
Fax: (64 4) 474 4111

7. The Chief Scientist (Pesticides)
Agricultural Compounds Unit
Ministry of Agriculture and Fisheries
P.O. Box 40-063
UPPER HUTT
Tel: (64 4) 528 6089
Fax: (64 4) 528 4675

8. The Codex Officer
Ministry of Agriculture
P.O. Box 2526
WELLINGTON
Tel: (64 4) 474 4100
Fax: (64 4) 474 4111

9. The Manager
Public Health Regulatory Services
Ministry of Health
P.O. Box 5013
WELLINGTON
Tel: (64 4) 496 2000
Fax: (64 4) 496 2340

10. Institute of Environmental Science and Research
Wellington Science Centre
P.O. Box 30-547
LOWER HUTT
Tel: (64 4) 570 1555
Fax: (64 4) 569 4500

Nicaragua
República de Nicaragua

Responsibility	
Main authority responsible for approval/registration of pesticides	1
Main lauthority responsible for pesticide use directives, permits and training	
Main authority responsible for establishing national maximum pesticide residue limits	1
Main authority responsible for control of pesticide residues in food	2, 3, 4, 5
Main laboratory for analysis of pesticide residues in food	4
Main laboratory for quality control of pesticide products	
Main organisation responsible for applied research regarding pesticide use. Information on related regional research	1
Designated national authority for notifications and information regarding banned and severely restricted pesticides including the PIC procedure	7
Codex Alimentarius contact point	8
National IRPTC correspondent	6
Infoterra national focal point	-
National IFCS focal point	10
National association of pesticide manufacturers	9

Responsible organisations

1. Ministerio de Agricultura y Ganaderia
Direccion General de Proteccion y Sanidad Agropecuaria
Km 3 1/2 Carretera a Masaya
MANAGUA
Tel: (502 2) 783412
Fax: (502 2) 785864

2. Laboratorio de Sanidad Animal
Direccion General de Proteccion y Sanidad Agropecuaria
Km 3 1/2 Carretera a Masaya
MANAGUA
Tel: (502 2) 784236
Fax: (502 2) 785864

3. Ministerio de Salud (MINSA)
MANAGUA

4. Laboratorio de Tecnologia de Alimentos (LABAL)
Ministerio de Industria
Apartado 8
MANAGUA

5. Direccion de Sanidad Animal
Direccion General de Proteccion y Sanidad Agropecuaria
MANAGUA
Tel: (502 2) 783417
Fax: (502 2) 785864

6. Especialista Ambiental del Proyecto Lago Xolotlan Biólogo
Instituto Nicaraguense de Recursos Naturales y del Ambiente (IRENA)
Km 12 1/2 Carretera Norte
Apartado 5123
MANAGUA D.N.

7. Director de Sanidad Vegetal
Dirección General de Protección y Sanidad Agropecuaria
Ministerio de Agricultura y Ganaderia
Km 3 1/2 Carretea a Masaya Apartado Postal 2335
2312 MANAGUA
Tel: (502 2) 784235
Fax: (502 2) 785 864, 780 224

8. Programa Normalización, Metrología y Control de Calidad (NMCC),
Dirección de Technología Industrial
Del Sandy's
Carretera Masaya, l c arriba
Apartado postal No. 8
MANAGUA

9. Asociacion Nicaraguense de Formuladores y Distribuidores de Agroquimicos (ANIFODA)
Del Hospital Velez Paiz
3 Cuadras Arriba,
Casa No 24
Apartado Postal 2067
MANAGUA
Tel: (505 2) 51516
Fax: (505 2) 51516

10. Dr. Jose' Luis Medina Bravo
Director General de Higiene y
Seguridad del Trabajo
Ministerio de Trabajo
MANAGUA
Tel: (505 2) 282 103
Fax: (505 2) 622 103

Niger
République du Niger

Responsibility	
Main authority responsible for approval/registration of pesticides	1
Main authority responsible for pesticide use directives, permits and training	
Main authority responsible for establishing national maximum pesticide residue limits	1
Main authority responsible for control of pesticide residues in food	-
Main laboratory for analysis of pesticide residues in food	2
Main laboratory for quality control of pesticide products	
Main organisation responsible for applied research regarding pesticide use. Information on related regional research	1, 3
Designated national authority for notifications and information regarding banned and severely restricted pesticides including the PIC procedure	1
Codex Alimentarius contact point	-
National IRPTC correspondent	-
Infoterra national focal point	4
National IFCS focal point	5
National association of pesticide manufacturers	-

Responsible organisations

1. Mr. le Directeur adjoint
   Direction de la Protection des Végétaux
   Ministère de l'Agriculture
   B.P. 323
   NIAMEY

2. Université de Niamey
   NIAMEY

3. Institut National de Recherche Agronomique (INRAN)
   NIAMEY

4. PFN Infoterra
   Attn: M. Algrabi Oussamane
   Division de la Documentation et de l'Archivage
   Service des Aménagements Forestiers
   B.P. 578 ou 12 520
   NIAMEY

5. Dr. Hamidou Beidari
   Inspecteur General des Pharmacies et Laboratoires du Niger
   Ministere de la Sante' Publique
   B.P. 511
   NIAMEY
   Fax: (227) 722424

Nigeria

Federal Republic of Nigeria

Responsibility	
Main authority responsible for approval/registration of pesticides	1, 7
Main authority responsible for pesticide use directives, permits and training	7
Main authority responsible for establishing national maximum pesticide residue limits	1 ,2, 7
Main authority responsible for control of pesticide residues in food	1, 2, 3, 7
Main laboratory for analysis of pesticide residues in food	1, 2, 7
Main laboratory for quality control of pesticide products	7
Main organisation responsible for applied research regarding pesticide use. Information on related regional research	4, 5, 7
Designated national authority for notifications and information regarding banned and severely restricted pesticides including the PIC procedure	6, 7
Codex Alimentarius contact point	8
National IRPTC correspondent	7
Infoterra national focal point	9
National IFCS focal point	7
National association of pesticide manufacturers	10

Note

Poisons information/control centre: Toxicology Unit, Department of Pharmacology, College of Medicine, University of Lagos, Tel: (234) 801500 [Head: Dr. A. Akintowa]

Responsible organisations

1. National Agency for Food and Drug Administration and Control
   Federal Secretariat, Phase II 2nd Floor
   P.M.B. 12525
   LAGOS
   Tel: 01-684 450

2. Standards Organisation of Nigeria
   P.M.B. 2102
   Yaba
   LAGOS
   Tel: 823 105, 823 772

3. Federal Ministry of Agriculture
   P.M.B. 12613
   LAGOS

4. Federal Ministry of Science and Technology
   P.M.B. 12793
   LAGOS

5. International Institute of Tropical Agriculture
   Moniya
   IBADAN

6. The Director
   Pest Control Services Division
   Federal Department of Livestock and Pest Control Services
   Area II, Garki
   P.M.B. 135
   ABUJA
   Tel: (09) 234 1329

7. The Director and Chief Executive,
   Federal Environmental Protection Agency
   Federal Secretariat, Phase II, 1st Floor
   P.M.B. 12620
   Ikoyi
   LAGOS
   Tel: (234 1) 585 1570, 680 071
   Fax: (234 1) 585 1571, 687 600

8. Prof. J.A Abalaka
   Director General,
   Standards Organisation of Nigeria
   Federal Secretariat
   Phase 1, 9th fl.
   P.M.B. 2102 (Yaba)
   Ikoyi
   LAGOS

9. Mr. Philip O. Bankole
   Infoterra NFP
   Federal Environmental Protection Agency
   The Presidency
   P.M.B. 265
   Garki
   ABUJA
   Tel: (234 9) 523 3368
   Fax: (234 9) 523 3373

10. Agrochemical Group of Manufacturers Association
    c/o CAPL
    Adeniyi Jones Avenue
    P.M.B. 21072
    IKEJA/LAGOS
    Tel: (234 1) 803 220
    Fax: (234 1) 684 182

Norway
Norge

Responsibility	
Main authority responsible for approval/registration of pesticides	1
Main authority responsible for pesticide use directives, permits and training	
Main authority responsible for establishing national maximum pesticide residue limits	2
Main authority responsible for control of pesticide residues in food	2
Main laboratory for analysis of pesticide residues in food	4
Main laboratory for quality control of pesticide products	
Main organisation responsible for applied research regarding pesticide use. Information on related regional research	4
Designated national authority for notifications and information regarding banned and severely restricted pesticides including the PIC procedure	1
Codex Alimentarius contact point	3
National IRPTC correspondent	7
Infoterra national focal point	6
National IFCS focal point	8
National association of pesticide manufacturers	5

Note

Poisons information/control centre: P.O. Box 8189 Dep., N-0034 OSLO, Tel: +47-260 8460, Fax: +47-260 8575 [Head: Dr (Mrs) E Wickström]

Responsible organisations

1. Norwegian Agricultural Inspection Service
   Pesticide Section
   P.O. Box 3
   N-1430 ÅS
   Tel: (47 64) 943 560
   Fax: (47 64) 942 350

2. Statens Nearingsmiddeltilsyn (SNT)
   P.O. Box 8187 Dep.
   N-0034 OSLO 1
   Tel: (47 22) 671 585
   Fax: (47 22) 199 533

3. Mr. John Race
   Norwegian Food Control Authority
   P.O. Box 8187
   N-0034 OSLO 1
   Tel: (47 22) 579 900
   Fax: (47 22) 579 901

4. Norwegian Crop Research Institute
   P.O. Box 100
   N-1430 ÅS
   Tel: 6494 2060
   Fax: 6494 2229

5. Norsk Plantevern Forening
   P.O. Box 53 Blindern
   N-0313 OSLO
   Tel: (47 22) 965 860, 940 316
   Fax: (47 22) 564 238, 550 180

6. Mr. Oyvind Saether
   Infoterra NFP
   Information Section
   Ministry of Environment
   P.O. Box 8013 Dep.
   N-0030 OSLO
   Tel: (47 22) 349 090
   Fax: (47 22) 342 758

7. The Product Control Department
   State Pollution Control Authority
   P.O. Box 8100 Dep.
   N-0032 OSLO 1
   Tel: (47 22) 573 400
   Fax: (47 22) 676 706

8. Mr. Joakim Lystad
   Head,
   Section for Industrial and Chemical Pollution
   Department for Water, Waste Management and Industry
   Ministry of Environmemt
   P.O. Box 8013 Dep.
   N-0030 OSLO
   Tel: (47 22) 349 090
   Fax: (47 22) 349 560

Oman
Saltanat 'Uman

Responsibility	
Main authority responsible for approval/registration of pesticides	1, 2, 3
Main authority responsible for pesticide use directives, permits and training	
Main authority responsible for establishing national maximum pesticide residue limits	-
Main authority responsible for control of pesticide residues in food	1, 4, 5
Main laboratory for analysis of pesticide residues in food	4, 5
Main laboratory for quality control of pesticide products	
Main organisation responsible for applied research regarding pesticide use. Information on related regional research	1
Designated national authority for notifications and information regarding banned and severely restricted pesticides including the PIC procedure	6
Codex Alimentarius contact point	7
National IRPTC correspondent	8, 11
Infoterra national focal point	9
National IFCS focal point	10
National association of pesticide manufacturers	-

Responsible organisations

1. Ministry of Agriculture & Fisheries
   Pesticide Control Committee
   P.O. Box 467
   MUSCAT

2. Council for Conservation of Environment and Prevention of Pollution
   P.O. Box 5310
   RUWI
   Tel: 704 344, 704 346

3. Dr Ahmed A K Al-Ghassany
   Director of Preventive Medicine
   Preventive Medicine Department
   Ministry of Health
   P.O. Box 393
   MUSCAT

4. Chemical and Bacteriological Laboratories
   Public Health Laboratories
   Directorate of Preventive Medicine
   Ministry of Health
   P.O. Box 393
   MUSCAT

5. Government Central Laboratory
   Ministry of Commerce and Industry
   P.O. Box 550
   MUSCAT

6. The Director of Agricultural Affairs
   Ministry of Agriculture and Fisheries
   MUSCAT
   Tel: 696 287
   Fax: 696 271

7. Director of Public Health
   Ministry of Health
   P.O. Box 393
   MUSCAT
   Tel: (968) 603 010
   Fax: (968) 601 430

8. The Director-General
   Health Affairs
   Ministry of Health
   P.O. Box 393
   MUSCAT 113

9. Mr. Mahmood Bin Mohammed Bin Talib Al-Zakwani
   Infoterra NFP
   Department of International Affairs
   Ministry of Regional Municipalities and Environment
   P.O. Box 323
   MUSCAT 113
   Tel: (968) 696 444
   Fax: (968) 602 320, 605 339

Responsible organisations

10. Mr A Mandhry
    Director,
    Environmental Health and
      Malaria Control
    Ministry of Health
    P.O. Box 393
    MUSCAT
    Tel: (968) 562 898
    Fax: (968) 563 121

11. The Director-General
    Health Control
    Ministry of Regional
      Municipality and
      Environment
    P.O. Box 461
    MUSCAT Code No 112

Pakistan

Islam-i Jamhuriya-e Pakistan

Responsibility	
Main authority responsible for approval/registration of pesticides	1
Main authority responsible for pesticide use directives, permits and training	
Main authority responsible for establishing national maximum pesticide residue limits	-
Main authority responsible for control of pesticide residues in food	-
Main laboratory for analysis of pesticide residues in food	3
Main laboratory for quality control of pesticide products	
Main organisation responsible for applied research regarding pesticide use. Information on related regional research	4, 5, 6, 7, 8, 9
Designated national authority for notifications and information regarding banned and severely restricted pesticides including the PIC procedure	10
Codex Alimentarius contact point	2
National IRPTC correspondent	12, 13
Infoterra national focal point	14
National IFCS focal point	-
National association of pesticide manufacturers	11

Note

Poisons information/control centre: Jinnah Post Graduate Medical Centre, 75510 Karachi, Emergency Tel: (92) 52 96 69 [Head: Dr T Matainaho]

Responsible organisations

1. Director and Adviser
   Plant Protection Department
   Government of Pakistan
   Ministry of Food and
     Agriculture
   Jinnah Avenue
   Malir Halt
   KARACHI

2. The Director General for
     Health
   Ministry of Health, Social
     Welfare and Population
     Planning
   Government of Pakistan
   Secretariat Block C
   ISLAMABAD
   Tel: (92 51) 820 930

3. Nutrition Division
   National Institute of Health
     P.O. NIH
   ISLAMABAD

4. Pakistan Agricultural Research
     Council
   Shahrah-e-Jumhuriat
   Ramna-5
   P.O. Box 1031
   ISLAMABAD

5. Vertebrate Pest Control
     Laboratory
   PARC
   Karachi Universitity Campus
   P.O. Box 8401
   KARACHI

6. Plant Protection Department
   Ministry of Food and
     Agriculture
   Block B
   Pakistan Secretariat
   ISLAMABAD

7. Pakistan Agricultural Research
     Council (PARC)
   L-13, Al-Markaz
   F-7 Commercial Area
   P.O. Box 1031
   ISLAMABAD

8. Ayub Agricultural Research
     Institutes
   Jhang Road
   FAISALABAD

9. National Agricultural
     Research Centre
   P.O. NIH
   Chakshahzad
   National Institute of Health
   ISLAMABAD

10. The Joint Secretary
    Environment and Urban
      Affairs Division
    Ministry of Housing and
      Works
    4th/5th Floor, Shaheed-e-
    Millat Secretariat, Blue Area
    ISLAMABAD
    Tel: (92 51) 822 757, 813 286

11. Pakistan Agricultural
      Pesticides Association
      (PAPA)
    c/o Ciba-Geigy Ltd
    No. 15 West Waref
    KARACHI
    Tel: (92 21) 200 112,
    441 517
    Fax: (92 21) 241 1009,
    441 517

12. The Vice-President
Pakistan Central Cotton Committee
Moulvi Tamizuddin Khan Road
KARACHI 1

13. The Director,
Pakistan Council of Scientific and Industrial Research (PCSIR)
Ferozepur Road
LAHORE 16

14. Dr. Muhammad Afzal
Director General,
Pakistan Scientific and Technological Information Centre (PASTIC)
P.O. Box 1217
ISLAMABAD
Tel: (92 51) 811 375, 824 161
Fax: (92 51) 811 381

Panama
República de Panamá

Responsibility	
Main authority responsible for approval/registration of pesticides	1, 2, 3, 4, 14
Main authority responsible for pesticide use directives, permits and training	
Main authority responsible for establishing national maximum pesticide residue limits	5, 6
Main authority responsible for control of pesticide residues in food	7, 8
Main laboratory for analysis of pesticide residues in food	7
Main laboratory for quality control of pesticide products	
Main organisation responsible for applied research regarding pesticide use. Information on related regional research	9
Designated national authority for notifications and information regarding banned and severely restricted pesticides including the PIC procedure	3
Codex Alimentarius contact point	11
National IRPTC correspondent	12
Infoterra national focal point	13
National IFCS focal point	-
National association of pesticide manufacturers	10

Responsible organisations

1. Ministerio de Salud
Departamento de Farmacia y Drogas
Apartado 2048
Zona 1
PANAMA CITY
Tel: (62) 6025, 3546
Fax: (62) 4354

2. Ministerio de Desarrollo Agropecuario
Dirección de Cuarentena Agropecuaria
Apartado 5390
Zona 5
PANAMA CITY
Tel: (32) 5340, 5358
Fax: (507) 325 340

3. Dirección de Sanidad Vegetal
Departamento de Registro y Control de Plaguicidas
Ministerio de Desarrollo Agropecuario
Apartado 5390,
Zona 5
PANAMA CITY
Tel: 207 979, 662 714
Fax: 662 012, 325 817

4. Ministerio de Salud
Departamento de Salud Ambiental
Apartado 2048,
Zona 1
PANAMA CITY
Tel: (62) 2180
Fax: (62) 4536

5. Ministerio de Salud
Departamento de Control de Alimentos y Vigilancia Veterinaria
Apartado 2048
Zona 1
PANAMA CITY
Tel: (62) 1902, 1619
Fax: (62) 4354

6. Ministerio de Comercio e Industrias
Dirección General de Normas y Tecnología Industrial
Apartado 9658,
Zona 4
PANAMA CITY
Tel: (27) 4749
Fax: (25) 7724

7. Laboratorio Especializado de Análisis
Universidad de Panamá
Estafa Universitaria
PANAMA CITY
Tel: (23) 6451
Fax: (69) 8880

8. Ministerio de Salud
Laboratorio Central de Salud
Apartado Postal 1474
Zona 1
PANAMA CITY
Tel: (62) 6521
Fax: (62) 6521

9. Instituto de Investigación Agropecuaria de Panama
Dirección Nacional de Servicio Cientifico 3
Apartado 6-4391
El Dorado 6A
PANAMA CITY
Tel: (64) 8702
Fax: (64) 9270

10. ANDIA
Edificio Transist - mica 1000
Oficina No. 107
Apartado Postal 476, Zona 9a
PANAMA CITY
Tel: (507) 611 264
Fax: (507) 610 906

11. Ing. Hersilia de Echervers
Directora General de Normas y Tecnología Industrial
Ministerio de Comercio e Industrias
Piso 19, Loteria Nacional de Beneficencia
Apartado 9658
Zona 4
PANAMA CITY

12. Jefe,
Sección de Farmacia y Drogas
Ministerio de Salud
Apartado 2048
Zona 1
PANAMA CITY

13. Sra. Maria Daysi de Gracia
Coordinadora del Centro Nacional de Infoterra
Comisión Nacional del Medio Ambiente
Ministerio de Planificación y Política Económica
Apartado 10120,
Zona 4
PANAMA CITY
Tel: (507) 326 055
Fax: (507) 696 822

14. Direccion Nacional de Sanidad Animal
Ministerio de Desarrollo Agropecuaria
Apartado 5390
PANAMA 5
Tel: (507) 266 1812, 220 7981
Fax: (507) 220 7981, 266 2943

Papua New Guinea

Independent State of Papua New Guinea

Responsibility	
Main authority responsible for approval/registration of pesticides	2
Main authority responsible for pesticide use directives, permits and training	2, 4
Main authority responsible for establishing national maximum pesticide residue limits	4, 7
Main authority responsible for control of pesticide residues in food	1, 3
Main laboratory for analysis of pesticide residues in food	3
Main laboratory for quality control of pesticide products	-
Main organisation responsible for applied research regarding pesticide use. Information on related regional research	3, 4
Designated national authority for notifications and information regarding banned and severely restricted pesticides including the PIC procedure	5
Codex Alimentarius contact point	6, 7
National IRPTC correspondent	5
Infoterra national focal point	5
National IFCS focal point	5
National association of pesticide manufacturers	-

Note

Poisons information/control centre: Department of Basic Medical Sciences, University of Papua New Guinea, Taurama Campus, P.O. Box 5623, Boroko NCD, Tel: (675) 248133, Fax: (675) 217658 [Head: Dr. T. Matainaho]

Responsible organisations

1. Department of Health
   P.O. Box 3991
   BOROKO, N.C.D.

2. Department of Environment and Conservation
   P.O. Box 6601
   BOROKO, N.C.D.

3. University of Papua New Guinea
   Biology Department
   P.O. Box 320
   The University
   WAIGANI, N.C.D.
   Tel: 245141

4. Department of Agriculture and Livestock
   P.O. Box 417
   KONEDOBU, N.C.D.

5. The Secretary,
   Department of Environment and Conservation
   Environment Division
   P.O. Box 6601
   BOROKO, N.C.D.
   Tel: (675) 327 1788, 271 275
   Fax: (675) 327 1900, 271 900

6. Dr. D.A. Watt
   Head,
   National Veterinary Laboratory
   Department of Agriculture and Livestock
   P.O. Box 6372
   BOROKO, N.C.D.

7. Director General
   National Institute of Standards and Industrial Technology
   P.O. Box 3042
   BOROKO, N.C.D.

Paraguay
República del Paraguay, Tetä Paraguáype

Responsibility	
Main authority responsible for approval/registration of pesticides	1
Main authority responsible for pesticide use directives, permits and training	
Main authority responsible for establishing national maximum pesticide residue limits	2
Main authority responsible for control of pesticide residues in food	2, 3
Main laboratory for analysis of pesticide residues in food	2
Main laboratory for quality control of pesticide products	
Main organisation responsible for applied research regarding pesticide use. Information on related regional research	1
Designated national authority for notifications and information regarding banned and severely restricted pesticides including the PIC procedure	4
Codex Alimentarius contact point	5
National IRPTC correspondent	-
Infoterra national focal point	6
National IFCS focal point	-
National association of pesticide manufacturers	7

Note

Poisons information/control centre: Catedra de Toxicologia, Universidad Nacional de Asuncion, Faculdad de Sciencias Medicas, Avenida Venezuela y Florida, Asuncion, Tel: (595) 21213734 [Head: Dr. E. Cardozo Melgazejo]

Responsible organisations

1. Ministerio de Agricultura y Ganadería
   Pte Franco 472
   ASUNCION
   Tel: 49404

2. Instituto Nacional de Tecnología y Normalización
   CC 967
   ASUNCION
   Tel: 290 160, 290 266, 290 277

3. Oficina Química Municipal
   Av. Gral Santos y Manuel Dominguez
   ASUNCION
   Tel: 203 084

4. Direccion de Defensa Vegetal
   Ayolas 177 Edif. Mercurio, 6 piso
   ASUNCION
   Tel: 440 307, 493 764
   Fax: 447 473, 440 306

5. Dr J Martino
   Director,
   Instituto Nacional de Tecnología y Normalización
   CC 967
   ASUNCION

6. Ing. For. Arnulfo Fretes Escario
   CNC de Infoterra
   Ministerio de Agricultura y Ganaderia
   Subsecretaria de Estado de Recursos Naturales y Medio Ambiente
   Tacuary No. 443, 4 piso
   ASUNCION
   Tel: (595 21) 443 971, 433 854
   Fax: (595 21) 440 167

7. Camara Paraguaya de Sanidad Agropecuaria y Fertilizantes
   (Address unknown)

Peru

República del Perú

Responsibility	
Main authority responsible for approval/registration of pesticides	1
Main authority responsible for pesticide use directives, permits and training	
Main authority responsible for establishing national maximum pesticide residue limits	1
Main authority responsible for control of pesticide residues in food	1, 2, 3, 4, 5, 6
Main laboratory for analysis of pesticide residues in food	1, 7, 8
Main laboratory for quality control of pesticide products	
Main organisation responsible for applied research regarding pesticide use. Information on related regional research	9, 10
Designated national authority for notifications and information regarding banned and severely restricted pesticides including the PIC procedure	11
Codex Alimentarius contact point	12
National IRPTC correspondent	13
Infoterra national focal point	14
National IFCS focal point	-
National association of pesticide manufacturers	15

Responsible organisations

1. Dirección General de Agricultura y Ganadería
Dirección de Sanidad Agricola
Av. Salaverry s/n
Jesus Maria
LIMA 11

2. Dirección General del Medio Ambiente (DIGEMA)
Pachacutec 900
LIMA 11

3. Ministerio de Agricultura
Av. Salaverry s/n
Edificio M. de Trabajo
LIMA 11

4. Laboratorio Químico Biológico
Pablo Bemudez 316
LIMA 11

5. Laboratorio Central de Control
Pasaje El Sol 777
Cdra. 34 Av. Argentina
CALLAO 3

6. Ministerio de Salud
Av. Salaverry Cdra. 8
LIMA 11

7. Instituto Nacional de Desarrollo Agroindustrial (INDDA)
Av. La Universidad 595
La Molina
LIMA 14

8 Instituto Nacional de Investigación y Promoción Agropecuaria (INIPA)
Guzman Blanco 309
LIMA 1

9. Instituto del Mar del Peru (IMARPE)
Gamarra y Gral. Valle s/n
Chucuito
CALLAO 1

10. Centro de Investigación en Nutrición y Control Alimentario (CINCA)
Tizon y Bueno 276
LIMA 11

11. Director General,
Dirección de Sanidad Vegetal
Servicio Nacional de Sanidad Agraria
Pasaje Zela S/N,
Jesus Maria
LIMA 11
Tel: (5114) 338 048
Fax: (5114) 338 048

12. Dr Carlos F Pastor
Director Ejecutivo,
Dirección de Higiene de Alimentos y Zoonosis
Direccion General de Salud Ambiental
Jr. las Amapolas No.350-Lince
LIMA
Tel: 428 353
Fax: 400 399

13. Director General,
Instituto Nacional de Recursos Naturales
Direccion General de Medio Ambiente
Apartado Postal 4452
Calle 17 No. 355
SAN ISIDRO

14. Ing. Miguel Ventura Napa
CNC de Infoterra
Institut Nacional de Recursos Naturales (INRENA)
Calle Diecisiete No. 355
Urb. El Palomar, San Isidro
Apartado Postal 4452
LIMA
Tel: (51 14) 410 425
Fax: (51 14) 414 606

15. Comité Importadores, Representantes y Fabricantes de Plaguicidas Agricolas de la Cámara de Comercio de Lima
Apartado 593
LIMA 11
Tel: (51 1) 463 3434
Fax: (51 1) 463 2820

Philippines
Republika ng Pilipinas, Republic of the Philippines

Responsibility	
Main authority responsible for approval/registration of pesticides	1
Main authority responsible for pesticide use directives, permits and training	
Main authority responsible for establishing national maximum pesticide residue limits	1
Main authority responsible for control of pesticide residues in food	1, 2
Main laboratory for analysis of pesticide residues in food	2, 3, 4
Main laboratory for quality control of pesticide products	
Main organisation responsible for applied research regarding pesticide use. Information on related regional research	5, 6, 7, 8, 9, 10
Designated national authority for notifications and information regarding banned and severely restricted pesticides including the PIC procedure	16
Codex Alimentarius contact point	11
National IRPTC correspondent	12
Infoterra national focal point	13
National IFCS focal point	14
National association of pesticide manufacturers	15

Note

Poisons information/control centre: College of Medicine, General Hospital, University of the Philippines, Taft Avenue, Ermita, Manila, Tel: (632) 521 8251, Fax: (632) 536 0062 [Head: Dr N Cortes-Maramba]

Responsible organisations

1. Fertilizer and Pesticide Authority (FPA)
   4th Floor, Bldg. B
   NIA Complex, EDSA
   P.O. Box 2582
   QUEZON CITY
   Tel: (63 2) 922 3368, 922 3371, 965 877
   Fax: (63 2) 922 3364

2. Bureau of Food and Drugs
   Food and Drug Administration DOH Compound
   Alabang, Muntinlupa
   METRO MANILA
   Tel: (63 2) 842 2213, 842 4625
   Fax: (63 2) 842 4603

3. Philippine Institute for Pure and Applied Chemistry
   Ateneo de Manila University
   Loyola Heights
   QUEZON CITY
   Tel: 924 4587

4. Bureau of Plant Industry (BPI)
   Pesticide Residue and Formulation Laboratory
   San Andrés
   MANILA
   Tel: 598540

5. Agricultural Information Bank of Asia (AIBA)
   College
   LAGUNA
   Tel: 2477

6. Department of Science and Technology
   Philippine Council for Agriculture, Resources and Development (PCARRD)
   Los Banos
   LAGUNA
   Tel: 50015

7. International Rice Research Institute
   P.O. Box 933
   Los Banos
   LAGUNA

8. Bureau of Plant Industry (BPI)
   Pesticide Residues and Formulation Laboratory
   692 San Andres Street
   Malate
   MANILA
   Tel: (632) 521 7650

9. National Crop Protection Center (NCPC)
   University of the Philippines at Los Banos College
   LAGUNA 3720
   Tel: 2409

10. Philippine Rice Research Institute
    UPLB Campus
    College
    LAGUNA
    Tel: (63 94) 3515, 3631
    Fax: (63 94) 3515

11. Office of the United Nations and Other International Organizations (UNIO)
Department of Foreign Affairs
2330 Roxas Boulevard
Pasay City
METRO MANILA
Tel: 834 4000
Fax: 832 3793, 832 5029

12. Environmental Management Bureau
Topaz Bldg,
99-101 Kamias Road
QUEZON CITY 3008
Tel: (63 2) 924 7540, 975 698
Fax: (63 2) 924 7540

13. Infoterra NFP
Technology and Livelihood Resource Center (TLRC)
Urban II Building
Sen. Gil J. Puyat Ave. Ext.
Makati 1200
METRO MANILA
Tel: (63 2) 859 811
Fax: (63 2) 818 7944

14. Dr Mario S Baquilod
Environmental Health Service
Department of Health
San Lazaro Compound
Rizal Avenue
Sta Cruz
METRO MANILA
Tel: (63 2) 711 6080
Fax: (63 2) 711 6050

15. Crop Protection Association of the Philippines (CPAP)
c/o Ciba-Geigy (Phils.) Inc.
Asian Reinsurance Building
Gamboa corner Salcedo Streets
Legaspi Village
MAKATI CITY
Tel: (63 2) 818 1372
Fax: (63 2) 872 644,
815 6263

16. The Director,
Fertilizer and Pesticide Authority (FPA)
Department of Agriculture
6th Floor, Raha Sulayman Building
Benavidez Street
Legaspi Village, Makati
METRO MANILLA
Tel: (63 2) 855 001, 818 1109
Fax: (62 2) 818 1363

Poland
Rzeczpospolita Polska

Responsibility

Main authority responsible for approval/registration of pesticides	1
Main authority responsible for pesticide use directives, permits and training	3
Main authority responsible for establishing national maximum pesticide residue limits	2
Main authority responsible for control of pesticide residues in food	4
Main laboratory for analysis of pesticide residues in food	5, 6
Main laboratory for quality control of pesticide products	7
Main organisation responsible for applied research regarding pesticide use. Information on related regional research	8
Designated national authority for notifications and information regarding banned and severely restricted pesticides including the PIC procedure	-
Codex Alimentarius contact point	10
National IRPTC correspondent	4
Infoterra national focal point	11
National IFCS focal point	2
National association of pesticide manufacturers	9

Note

Poisons information/control centres in Gdansk, Krakow, Lodz, Lublin, Poznan, Rzeszow, Sosnowiec, Warsaw and Wroclaw

Responsible organisations

1. Ministry of Agriculture and Food Economy
Department of Plant Production
ul. Wspolna 30
00-930 WARSAW
Tel: (48 22) 623 2462
Fax: (48 22) 628 8784

2. Ministry of Health and Social Welfare
Department of Public Health
ul. Miodowa 15
00-923 WARSAW
Tel: (48 22) 261 521
Fax: (48 22) 635 8852

3. Plant Protection Institute
ul.Miczurina 20
60-318 POZNAN
Tel: (48 61) 679 021
Fax: (48 61) 676 301

4. National Institute of Hygiene
Deparment of Environmental Toxicology
ul. Chocimska 24
00-791 WARSAW
Tel: (48 22) 497 084
Fax: (48 22) 497 484

5. Plant Protection Institute
Department of Plant Protection Products Residue Analyses
ul. Miczurina 20
60-318 POZNAN
Tel: (48 61) 674 841
Fax: (48 61) 676 301

6. Veterinary Institute
Department of Pharmacology and Toxicology
ul. Partyzantow 57
24-100 PULAWY
Tel: (48 86) 3051
Fax: (48 86) 2595

7. Plant Protection Institute
Research Branch in Sosnicowice
Department of Plant Protection Products Analyses
ul. Gliwicka 29
44-153 SOSNICOWICE
Tel: (48 32) 387 584
Fax: (48 32) 387 503

8. Institute of Plant Protection
Office for Plant Protection Products Registration
ul. Miczurina 20
60-318 POZNAN
Tel: (48 61) 675 713
Fax: (48 61) 676 301

9. Institute of Organic Chemistry
ul. Annopol 6
03-236 WARSAW
Tel: (48 22) 111 231
Fax: (48 22) 110 700

10. Quality Inspection Office
ul. Zurawia 32/34
00-950 WARSAW
Tel: (48 22) 621 6421
Fax (48 22) 621 4858

Responsible organisations

11. Institute of Environmental
    Protection
    ul. Krucza 5/11
    00-548 WARSZAW
    Tel: (48 22) 628 1569
    Fax (48 22) 295 263

Portugal

República Portuguesa

Responsibility	
Main authority responsible for approval/registration of pesticides	1, 2, 3
Main authority responsible for pesticide use directives, permits and training	
Main authority responsible for establishing national maximum pesticide residue limits	1
Main authority responsible for control of pesticide residues in food	2, 4, 5
Main laboratory for analysis of pesticide residues in food	2, 5
Main laboratory for quality control of pesticide products	
Main organisation responsible for applied research regarding pesticide use. Information on related regional research	2, 8, 10
Designated national authority for notifications and information regarding banned and severely restricted pesticides including the PIC procedure	11
Codex Alimentarius contact point	12
National IRPTC correspondent	13
Infoterra national focal point	9
National IFCS focal point	7
National association of pesticide manufacturers	6

Note

Poisons information/control centre: Centro de Informacao Antivenenos, Instituto Nacional de Emergencia Medica, Rua Infante D. Pedro 8, 1799 Lisboa Cedex, Tel: (351 1) 7930503, Fax: (351 1) 7937124 [Head: Dr. Ms. A. Borges]

Responsible organisations

1. Comissáo de Toxicologia de Pesticides
   Quinta do Marqués
   P-2780 OEIRAS

2. Direccáo-Geral de Proteccáo da Producáo Agricola
   Quinta do Marqués
   P-2780 OEIRAS
   Tel: 2430772

3. Direccáo-Geral do Comércio náo Alimentar
   Av. Visconde Valmor 72
   P-1000 LISBOA

4. Instituto da Qualidade Alimentar (IQA)
   Rua Sociedade Farmacéutica 39
   P-1100 LISBOA

5. Instituto Nacional de Saúde
   Av. Paulo VI
   P-1700 LISBOA

6. Associao National da Industria para Proteccao das Plantes (ANIPLA)
   Rua Dos Navegantes
   48 r/c dt
   P-1200 LISBOA
   Tel: (351 1) 601 051,
   395 0828
   Fax: (351 1) 601 292,
   395 0910

7. Dr. M. Pinho da Silva
   Divisao de Saude Ambiental
   Direccao-Geral de Saude
   Al. D. Alfonso Henriques, 45
   P-1056 LISBOA CEDEX
   Tel: (351 1) 847 5515
   Fax: (351 1) 847 6639,
   847 6455

8. Instituto Superior de Agronomia
   Tapada da Ajuda
   P-1300 LISBOA

9. Sr. A. L. Madureira
   Infoterra NFP
   Secretaria-General
   Ministerio do Ambiente e Recursos Naturais
   Rua de O. Seculo, 51
   1200 LISBOA
   Tel: (351 1) 346 2751
   Fax: (351 1) 576 774

10. Estacáo Agrónima Nacional
    Quinta do Marqués
    P-2780 OEIRAS

11. The Director
    Centro Nacional da Protecao da Producao Agrícola
    Quinta do Marques
    P-2780 OEIRAS
    Tel: (1) 441 2822
    Fax: (1) 442 0616

12. Sub-Commissáo do Codex Alimentarius
    Commissáo Nacional da FAO
    Ministério dos Negócios Estrangeiros
    Pal. das Necessidades
    Largo do Rilvas
    P-1354 LISBOA CODEX

13. Direccáo-Geral de Qualidade do Ambiente
    Secretária de Estado do Ambiente
    Rua do Seculo 51-20
    P-1200 LISBOA

Qatar

Dawlat Qatar, State of Qatar

Responsibility	
Main authority responsible for approval/registration of pesticides	1, 3
Main authority responsible for pesticide use directives, permits and training	3, 5
Main authority responsible for establishing national maximum pesticide residue limits	2
Main authority responsible for control of pesticide residues in food	1, 2
Main laboratory for analysis of pesticide residues in food	2
Main laboratory for quality control of pesticide products	2
Main organisation responsible for applied research regarding pesticide use. Information on related regional research	3, 4
Designated national authority for notifications and information regarding banned and severely restricted pesticides including the PIC procedure	5
Codex Alimentarius contact point	6
National IRPTC correspondent	7
Infoterra national focal point	8
National IFCS focal point	1
National association of pesticide manufacturers	-

Responsible organisations

1. Ministry of Public Health
   P.O. Box 42
   DOHA
   Tel: 441 555

2. Regional Centre for Food Contamination Monitoring
   P.O. Box 7083
   DOHA
   Tel: 417 676

3. Ministry of Municipal Affairs and Agriculture
   DOHA
   Tel: 336 336

4. Scientific and Applied Research Centre
   University of Qatar
   P.O. Box 2713
   DOHA
   Tel: 874 961

5. The Chief,
   Crop Protection and Quarantine
   Ministry of Agriculture
   P.O. Box 1966
   DOHA
   Tel: 421 249, 433 400
   Fax: 436 977, 322 002

6. The Under-Secretary
   Ministry of Public Health
   P.O. Box 3050
   Tel: (974) 352 324
   Fax: (974) 433 019

7. The Director,
   Medical and Public Health Services
   Ministry of Public Health
   P.O. Box 42
   DOHA

8. Mr. Khalid Ghanim Al Ali
   Infoterra NFP
   Environment Department
   Ministry of Municipal Affairs and Agriculture
   P.O. Box 7634
   DOHA
   Tel: (974) 437 789
   Fax: (974) 415 246

Romania
România

Responsibility

Main authority responsible for approval/registration of pesticides	7
Main authority responsible for pesticide use directives, permits and training	
Main authority responsible for establishing national maximum pesticide residue limits	1
Main authority responsible for control of pesticide residues in food	2
Main laboratory for analysis of pesticide residues in food	2
Main laboratory for quality control of pesticide products	
Main organisation responsible for applied research regarding pesticide use. Information on related regional research	1
Designated national authority for notifications and information regarding banned and severely restricted pesticides including the PIC procedure	3
Codex Alimentarius contact point	4
National IRPTC correspondent	5
Infoterra national focal point	6
National IFCS focal point	-
National association of pesticide manufacturers	-

Responsible organisations

1. Research Institute for Plant Protection
   24 Nicolae Bâlcescu Avenue
   BUCHAREST
   Tel: 144 020

2. Central Control Laboratory for Food of Animal Origin
   24 Nicolae Bâlcescu Avenue
   BUCHAREST
   Tel: 144 020

3. The Director
   General Division for Strategy Legislation and Ecological Reconstruction
   Ministry of Environment
   12 Bd. Libertatii, Tronson 5
   BUCHAREST
   Tel: (40 0) 316 146
   Fax: (40 0) 120 403

4. Institut Roumain de Normalisation
   13, Jean Louis Calderon
   Code 70201
   BUCHAREST 2
   Tel: (40 0) 114 043
   Fax: (40 0) 120 823

5. The Director
   Monitoring Division
   Attn: Mr. Dumitru Mihu
   Ministry of Environment
   Bd. Libertatji No. 12, sector 5
   BUCHAREST
   Tel: (40 0) 316 117
   Fax: (40 0) 316 486

6. Directeur Sanda Odiatiu
   Office d'Information et de Documentation pour l'Environnement
   Bd. Libertatii, No. 14,
   sector 5
   BUCHAREST
   Tel: (40 1) 781 7877
   Fax: (40 1) 311 3308

7. Ministry of Agriculture and Food Industry
   Veterinary Department
   24, Nicolae Balcescu Avenue
   BUCHAREST
   Tel: 144 020

Russian Federation
Rossiyskaya Federatsiya

Responsibility	
Main authority responsible for approval/registration of pesticides	1, 2, 3, 5
Main authority responsible for pesticide use directives, permits and training	1
Main authority responsible for establishing national maximum pesticide residue limits	3, 4, 5
Main authority responsible for control of pesticide residues in food	
Main laboratory for analysis of pesticide residues in food	
Main laboratory for quality control of pesticide products	
Main organisation responsible for applied research regarding pesticide use. Information on related regional research	7, 8, 9
Designated national authority for notifications and information regarding banned and severely restricted pesticides including the PIC procedure	2, 6
Codex Alimentarius contact point	
National IRPTC correspondent	6
Infoterra national focal point	10
National IFCS focal point	3
National association of pesticide manufacturers	11

Note

Poisons information/control centre: INTOCENTER, Soukharevskaya Square 3, Block 7, 129 010 Moskva, Tel: (95) 978 75 41 [Head: Dr Ju N Ostapenko]

Responsible organisations

1. State Commission for Chemical Control of Plant Diseases and Weeds (Goskhimkommissiya)
   Orlikov per. 1/11
   107139 MOSCOW
   Tel: (095) 207 6390, 208 6284
   Fax: (095) 207 4233

2. Ministry of Agriculture and Foodstuffs (Minselkhozprod)
   Department for Chemical Protection
   Orlikov per. 1/11
   107139 MOSCOW
   Tel: (095) 975 6367
   Fax: (095) 207 6306

3. State Committee for Sanitary and Epidemiological Surveillance (Goskomsanepidnadzor)
   Vadkovsky per. 18/20
   101479 MOSCOW
   Tel: (095) 973 1717, 973 2674
   Fax: (095) 973 1398

4. State Committee for Sanitary and Epidemiological Surveillance (Goskomsanepidnadzor)
   Department for Sanitary and Epidemiological Surveillance and Expertise
   Vadkovsky per. 18/20
   101479 MOSCOW
   Tel: (095) 973 1395
   Fax: (095) 975 1549

5. Centre for Hygienic Regulation of Agricultural Chemicals
   Erisman Scientific Research Institute for Hygiene
   Semashko 2
   Mytishchi
   141000 MOSCOW
   Tel: (095) 582 9698, 586 1072, 586 1144
   Fax: (095) 582 9294

6. Russian Register of Potentially Hazardous Chemical and Biological Substances
   Vadkovsky per. 18/20
   101479 MOSCOW
   Tel:(095) 973 2657, 973 3021
   Fax: (095) 973 2657

7. All-Russian Research Institute for Plant Protection (VIZR)
   Podbelskyi Chaussee 3
   Pushkin
   ST. PETERSBURG
   Tel: (812) 470 5110, 470 5129, 470 5133
   Fax: (812) 476 4388

8. Research Centre for Toxicological and Hygienic Regulation of Biopreparations
   Lenin Str. 102A
   Dashkovka
   142283 MOSCOW

Responsible organisations

9. All-Russian Research Institute for Veterinary Sanitary, Hygiene and Ecology (VNIIVSGE)
   Zvenigorodskoye Chaussee 5
   123022 MOSCOW
   Tel: (095) 522 2905, 522 2920

10. Professor N.G. Rybalsky
    Infoterra NFP
    Director-General,
    Russian Ecological
    Information Agency
    Ministry of Environment and Natural Resources
    B. Gruzinskaya Str.
    123812 MOSCOW
    Tel: (7095) 284 8235
    Fax: (7095) 284 8550

11. Agroprochim - Industrial Association for Plant Protection Chemicals
    Ugreshskaya, 31
    109088 MOSCOW
    Tel: 277 6472

Rwanda

Repubulika y'u Rwanda, République Rwandaise

Responsibility	
Main authority responsible for approval/registration of pesticides	1
Main authority responsible for pesticide use directives, permits and training	
Main authority responsible for establishing national maximum pesticide residue limits	-
Main authority responsible for control of pesticide residues in food	-
Main laboratory for analysis of pesticide residues in food	-
Main laboratory for quality control of pesticide products	-
Main organisation responsible for applied research regarding pesticide use. Information on related regional research	-
Designated national authority for notifications and information regarding banned and severely restricted pesticides including the PIC procedure	1
Codex Alimentarius contact point	2
National IRPTC correspondent	3
Infoterra national focal point	4
National IFCS focal point	-
National association of pesticide manufacturers	

Responsible organisations

1. Monsieur le Directeur
   Direction de la Protection des Végétaux
   Ministère de l'Agriculture, de l'Elevage et des Fôrets
   B.P. 621
   KIGALI
   Tel: 73415
   Fax: 75689

2. Division Normalisation et Contrôle de la Qualité
   Ministère du Commerce et de la Consommation
   B.P 476
   KIGALI
   Tel: (250) 73875, 73237
   Fax: (250) 72984

3. Monsieur le Directeur
   Projet environnement et developpement
   Ministere du plan
   B.P 46
   KIGALI

4. Mr. Jean Nepomucene Nayinzira
   Directeur,
   PFN Infoterra
   Ministère de l'Environnement et du Tourisme
   B.P. 2378
   KIGALI
   Tel: (250) 77454
   Fax: (250) 76958

Saint Kitts and Nevis

Federation of Saint Kitts and Nevis

Responsibility	
Main authority responsible for approval/registration of pesticides	1
Main authority responsible for pesticide use directives, permits and training	1
Main authority responsible for establishing national maximum pesticide residue limits	
Main authority responsible for control of pesticide residues in food	-
Main laboratory for analysis of pesticide residues in food	-
Main laboratory for quality control of pesticide products	-
Main organisation responsible for applied research regarding pesticide use. Information on related regional research	1, 3
Designated national authority for notifications and information regarding banned and severely restricted pesticides including the PIC procedure	1
Codex Alimentarius contact point	-
National IRPTC correspondent	-
Infoterra national focal point	2
National IFCS focal point	-
National association of pesticide manufacturers	

Responsible organisations

1. The Director of Agriculture
   Pesticide Control Board
   Ministry of Agriculture,
   Lands, Housing and
   Development
   Government Headquarters
   P.O. Box 186
   ST. KITTS
   Tel: (1 809) 465 2335

2. Mr. Leonard Huggins
   Infoterra NFP
   Department of Planning
   CHARLESTOWN, Nevis
   Tel: (1 809) 469 5521,
   ext. 2040
   Fax: (1 809) 469 5485

3. Mr. S Weekes
   Caribbean Agricultural and
   Development Institute
   (CARDI)
   Agriculture Department
   ST. KITTS
   Tel: (1 809) 465 4826
   Fax: (1 809) 465 1498

Saint Lucia

Responsibility	
Main authority responsible for approval/registration of pesticides	1
Main authority responsible for pesticide use directives, permits and training	2
Main authority responsible for establishing national maximum pesticide residue limits	-
Main authority responsible for control of pesticide residues in food	2, 3
Main laboratory for analysis of pesticide residues in food	3
Main laboratory for quality control of pesticide products	-
Main organisation responsible for applied research regarding pesticide use. Information on related regional research	4, 5
Designated national authority for notifications and information regarding banned and severely restricted pesticides including the PIC procedure	1
Codex Alimentarius contact point	3
National IRPTC correspondent	-
Infoterra national focal point	6
National IFCS focal point	-
National association of pesticide manufacturers	-

Responsible organisations

1. National Pesticides Control Board
   Ministry of Agriculture, Lands, Fisheries and Forestry
   Manoel Street
   CASTRIES
   Tel: (809) 452 2526
   Fax: (809) 453 6314

2. Ministry of Agriculture
   NIS Building
   CASTRIES
   Tel: (809) 452 2526
   Fax: (809) 453 6314

3. Produce Chemist Laboratory
   Union Agricultural Station
   Ministry of Agriculture Union
   CASTRIES
   Tel: (809) 450 2375
   Fax: (809) 450 1185

4. Research and Development Division
   Crop Protection Unit
   Ministry of Agriculture Union
   CASTRIES
   Tel: (809) 450 23375
   Fax: (809) 450 1185

5. Windward Island Banana Association (WINBAN)
   P.O. Box 115
   CASTRIES
   Tel: (809) 451 4242
   Fax: (809) 451 4601

6. Mrs. Vanesta Moses-Felix
   Infoterra NFP
   National Documentation Centre
   Ministry of Planning, Development and Environment
   P.O. Box 709
   CASTRIES
   Tel: (809) 453 2005, 453 1951
   Fax: (809) 452 2506

Saint Vincent and the Grenadines

Responsibility

Main authority responsible for approval/registration of pesticides	
Main authority responsible for pesticide use directives, permits and training	
Main authority responsible for establishing national maximum pesticide residue limits	
Main authority responsible for food control of pesticide residues in food	
Main laboratory for analysis of pesticide residues in food	
Main laboratory for quality control of pesticide products	
Main organisation responsible for applied research regarding pesticide use. Information on related regional research	
Designated national authority for notifications and information regarding banned and severely restricted pesticides including the PIC procedure	1
Codex Alimentarius contact point	-
National IRPTC correspondent	-
Infoterra national focal point	-
National IFCS focal point	-
National association of pesticide manufacturers	

Responsible organisations

1. The Chief Agricultural Officer
   Ministry of Agriculture,
     Industry and Labour
   KINGSTOWN
   Tel: (809 452) 61021
   Fax: (809 45) 62873

Sao Tomé and Príncipe

República Democrática de Sao Tomé e Príncipe

Responsibility	
Main authority responsible for approval/registration of pesticides	
Main authority responsible for pesticide use directives, permits and training	
Main authority responsible for establishing national maximum pesticide residue limits	
Main authority responsible for control of pesticide residues in food	
Main laboratory for analysis of pesticide residues in food	
Main laboratory for quality control of pesticide products	
Main organisation responsible for applied research regarding pesticide use. Information on related regional research	
Designated national authority for notifications and information regarding banned and severely restricted pesticides including the PIC procedure	1
Codex Alimentarius contact point	-
National IRPTC correspondent	-
Infoterra national focal point	2
National IFCS focal point	-
National association of pesticide manufacturers	

Responsible organisations

1. Monsieur le Directeur
   Direction de l'aménagement du territoire et de l'environnement
   Ministère de l'equipment social et de l'environnement
   C.P. No 67
   SAO TOME
   Fax: 22936

2. Mr. Joao da Conceicao
   PFN d'Infoterra
   Direccao de Ordenamento do Territorio e Meio Ambiente
   C.P. 67
   SAO TOME
   Tel: 22308

Saudi Arabia
al-Mamlakah al-'Arabíyah as-Saúdiyah

Responsibility	
Main authority responsible for approval/registration of pesticides	1, 2
Main authority responsible for pesticide use directives, permits and training	
Main authority responsible for establishing national maximum pesticide residue limits	1
Main authority responsible for control of pesticide residues in food	1
Main laboratory for analysis of pesticide residues in food	1, 4
Main laboratory for quality control of pesticide products	
Main organisation responsible for applied research regarding pesticide use. Information on related regional research	1
Designated national authority for notifications and information regarding banned and severely restricted pesticides including the PIC procedure	3
Codex Alimentarius contact point	7
National IRPTC correspondent	5
Infoterra national focal point	6
National IFCS focal point	8, 9
National association of pesticide manufacturers	-

Responsible organisations

1. Ministry of Agriculture and Water (MAW)
   11195 RIYADH
   Tel: (966 1) 401 6666
   Fax: (966 1) 405 5848

2. Ministry of Municipalities and Rural Affairs (MOMRA)
   11136 RIAYDH
   Tel: (966 1) 441 8888
   Fax: (966 1) 441 7222

3. The President
   Meteorological and Environmental Protection Administration (MEPA)
   Ministry of Defense and Aviation
   P.O. Box 1358
   21431 JEDDAH
   Tel: (966 2) 651 8676
   Fax: (966 2) 651 1424

4. Food Quality Control Laboratory
   Ministry of Commerce
   11162 RIYADH
   Tel: (966 1) 401 2222
   Fax: (966 1) 403 8421

5. The Director,
   Division of Environmental Standards,
   Meteorology and Environmental Protection Administration (MEPA)
   P.O. Box 1358
   21431 JEDDAH
   Fax: (966 2) 651 1424

6. The Director-General,
   National Meteorology and Environmental Protection Administration (MEPA)
   Attn: Dr Islem A Basunbul
   P.O. Box 1358
   21431 JEDDAH
   Tel: (966 2) 653 1672
   Fax: (966 2) 651 1424

7. International Relations Department
   Saudi Arabian Standards Organization (SASO)
   P.O. Box 3437
   11471 RIYADH
   Tel: 452 0194
   Fax: 452 0147

8. Mr. Sulaiman Mohamed Al-Zebn
   Director,
   National Program for Chemical Safety
   Biometeorology and Environmental Protection Agency
   Ministry of Defence and Aviation
   P.O. Box 1358
   21431 JEDDAH
   Tel: 651 7832
   Fax: 651 7832

9. Mr. Awad Oweid Al-Khattabi
   Director General,
   International Health Department
   Ministry of Health
   11176 RIYADH
   Tel: 403 4381

Senegal
République du Sénégal

Responsibility	
Main authority responsible for approval/registration of pesticides	1, 2, 3, 4, 5
Main authority responsible for pesticide use directives, permits and training	
Main authority responsible for establishing national maximum pesticide residue limits	1, 2, 3, 4, 6
Main authority responsible for control of pesticide residues in food	1
Main laboratory for analysis of pesticide residues in food	1, 2, 5
Main laboratory for quality control of pesticide products	
Main organisation responsible for applied research regarding pesticide use. Information on related regional research	1, 2, 3
Designated national authority for notifications and information regarding banned and severely restricted pesticides including the PIC procedure	7
Codex Alimentarius contact point	8
National IRPTC correspondent	9
Infoterra national focal point	9
National IFCS focal point	-
National association of pesticide manufacturers	-

Responsible organisations

1. L'Institut de Technologie Alimentaire (ITA)
   B.P. 2765
   DAKAR

2. L'École Inter-États des Sciences et Médicine Vétérinaire (EISMV)
   Université de Dakar
   DAKAR

3. L'Institut Sénégalais de Recherches Agricoles (ISRA)
   Ministère du Développement Rurale
   DAKAR

4. Le Service de l'Alimentation et de la Nutrition appliquée au Sénégal (SANAS)
   Ministère de la Santé Publique
   DAKAR

5. Faculté de Médecine et de Pharmacie
   Université de Dakar
   DAKAR

6. L'Institut sénégalais de Normalisation (ISN)
   Ministère de Plan et de la Cooperation
   DAKAR

7. Monsieur le Directeur
   Direction de l'environnement
   Ministère de la protection de la nature
   104 rue Carnot
   DAKAR
   Tel: (221) 219 052

8. Comité National du Codex
   Service de l'Alimentation et de la Nutrition Appliquée au Sénégal (SANAS)
   Ministère de la Santé Publique
   DAKAR

9. M. Mohamed Fadhel Diagne
   PFN d'Infoterra
   Centre National de Documentation Scientifique et Technique (CNDST)
   Ministère de la Modernisation de l'État et de la Technologie
   B.P. 3218
   DAKAR
   Tel: (221) 215 163
   Fax: (221) 229 764

Seychelles
Republic of Seychelles, Repiblik Sesel, République des Seychelles

Responsibility	
Main authority responsible for approval/registration of pesticides	1
Main authority responsible for pesticide use directives, permits and training	1
Main authority responsible for establishing national maximum pesticide residue limits	1
Main authority responsible for control of pesticide residues in food	2
Main laboratory for analysis of pesticide residues in food	3
Main laboratory for quality control of pesticide products	4
Main organisation responsible for applied research regarding pesticide use. Information on related regional research	2, 5
Designated national authority for notifications and information regarding banned and severely restricted pesticides including the PIC procedure	1
Codex Alimentarius contact point	2, 4
National IRPTC correspondent	6
Infoterra national focal point	-
National IFCS focal point	-
National association of pesticide manufacturers	-

Responsible organisations

1. Pesticides Board
Ministry of Health
P.O. Box 52
VICTORIA (Mahé)
Tel: (248) 388000
Fax: (248) 224792

2. Environmental
Health Section
Ministry of Health
P.O. Box 52
VICTORIA (Mahé)
Tel: (248)224400
Fax: (248)224792, 225416

3. Public Health Laboratory
Ministry of Health
P.O. Box 52
VICTORIA (Mahé)
Tel: (248)224400
Fax: (248)224792, 225416

4. Seychelles Bureau of Standards
P.O. Box 953
VICTORIA (Mahé)
Tel: (248) 375333
Fax: (248) 375151

5. Plant Protection Section
Ministry of Agriculture and
Marine Resources
P.O. Box 166
VICTORIA (Mahé)
Tel: (248) 322411
Fax: (248) 225245

6. Director General
Disease Prevention and
Control
Ministry of Health
P.O. Box 52
VICTORIA (Mahé)

Sierra Leone
Republic of Sierra Leone

Responsibility	
Main authority responsible for approval/registration of pesticides	1
Main authority responsible for pesticide use directives, permits and training	
Main authority responsible for establishing national maximum pesticide residue limits	2, 3
Main authority responsible for control of pesticide residues in food	4
Main laboratory for analysis of pesticide residues in food	4, 5, 6
Main laboratory for quality control of pesticide products	
Main organisation responsible for applied research regarding pesticide use. Information on related regional research	7, 8, 9
Designated national authority for notifications and information regarding banned and severely restricted pesticides including the PIC procedure	10
Codex Alimentarius contact point	11
National IRPTC correspondent	12
Infoterra national focal point	13
National IFCS focal point	-
National association of pesticide manufacturers	-

Responsible organisations

1. Phytosanitary Division
Ministry of Agriculture and Forestry
Tower Hill
FREETOWN

2. Ministry of Agriculture and Forestry
Tower Hill
FREETOWN

3. Ministry of Health
Gloucester Street
FREETOWN

4. Analytical Chemistry Laboratory
Central Laboratories
Connaught Hospital
FREETOWN

5. Nutrition Department
Njala University College
NJALA

6. Chemistry Department
Fourah Bay College
Mount Aureol
FREETOWN

7. Department of Botany
Njala University
NJALA

8. Department of Botany
Fourah Bay College
Mount Aureol
FREETOWN

9. Rice Research Station
ROKUPUR

10. The Principal Crop Protection Officer
Room 102W, MANR&F
FREETOWN

11. Mr A B Turay
Chief of Standards Central
Contact Codex Alimentarius Commission
National Bureau of Standards
Ministry of Trade and Industry
George Street
FREETOWN

12. The Director,
Government Medical Stores
Ministry of Health
New England
FREETOWN

13. Mr. Ibrahim M. Kamara
Infoterra NFP
c/o The Director,
Land and Water Development Division
P.O. Box 187
Tower Hill
FREETOWN
Tel: (232 22) 222621

Singapore

Hsin-chia-p'o Kung-ho-kuo, Republik Singapura, Singapore Kudiyarasu, Republic of Singapore

Responsibility	
Main authority responsible for approval/registration of pesticides.	1, 6
Main authority responsible for pesticide use directives, permits and training	1, 6
Main authority responsible for establishing national maximum pesticide residue limits	3
Main authority responsible for control of pesticide residues in food.	3
Main laboratory for analysis of pesticide residues in food	2, 6
Main laboratory for quality control of pesticide products	6
Main organisation responsible for applied research regarding pesticide use. Information on related regional research.	2, 6
Designated national authority for notifications and information regarding banned and severely restricted pesticides including the PIC procedure	-
Codex Alimentarius contact point	3
National IRPTC correspondent	4
Infoterra national focal point	5
National IFCS focal point	-
National association of pesticide manufacturers	-

Note

Poisons information/control centre: Institute of Science and Forensic Medicine, National Blood Centre, Toutram Road, Singapore 0316, Tel: (65) 223 5454 [Head: Dr. B. Ch. Bloodworth]

Responsible organisations

1. Ministry of the Environment
   Pollution Control Department
   12th Floor, Environment Building
   40 Scotts Road
   SINGAPORE 228231
   Tel: 65-732 7733
   Fax: 65-731 9651

2. Department of Scientific Services
   Institute of Science and Forensic Medicine
   1 Second Hospital Avenue
   SINGAPORE 168935
   Tel: 65-221 6800
   Fax: 65-229 0749

3. Ministry of the Environment
   Food Control Department
   Environment Building
   40 Scotts Road
   SINGAPORE 228231
   Tel: 65-731 9819
   Fax: 65-731 9843

4. The Deputy Director (Resources)
   Department of Scientific Services
   Institute of Science and Forensic Medicine
   1 Second Hospital Avenue
   SINGAPORE 168935
   Tel: 65-229 0720
   Fax: 65-229 0749

5. Ms. Fauziah Soeratman
   Librarian,
   Environment Resource Centre
   Ministry of the Environment
   6th Floor, Environment Building
   40 Scotts Road
   SINGAPORE 228231
   Tel: 65-731 9955
   Fax: 65-731 9456

6. Ministry of National Development
   Primary Production Department
   5 Maxwell Road
   #02-00/#03-00
   Tower Block, MND Complex
   SINGAPORE 069110
   Tel: 65-222 1211
   Fax: 65-220 6068

Slovakia

Slovenská Republika, Slovak Republic

Responsibility	
Main authority responsible for approval/registration of pesticides	1, 2
Main authority responsible for pesticide use directives, permits and training	2
Main authority responsible for establishing national maximum pesticide residue limits	1, 3
Main authority responsible for control of pesticide residues in food	4
Main laboratory for analysis of pesticide residues in food	3, 5, 6
Main laboratory for quality control of pesticide products	2
Main organisation responsible for applied research regarding pesticide use. Information on related regional research	1
Designated national authority for notifications and information regarding banned and severely restricted pesticides including the PIC procedure	2
Codex Alimentarius contact point	7
National IRPTC correspondent	8
Infoterra national focal point	9
National IFCS focal point	10
National association of pesticide manufacturers	-

Note

Poisons information/control centre: Clinic of Occupational Diseases, Dumbierska 3, Krasna Horka, 83101 Bratislava, Tel: (42) 371061 [Head: Mr. I. Batora]

Responsible organisations

1. Ministry of Soil Management
   Dobrovicova 12
   81266 BRATISLAVA

2. Central Agricultural Control and Testing Institute - UKSUP
   Ministry of Agriculture
   Matuskova 21
   83316 BRATISLAVA
   Tel: (42 7) 375 822, 375 861
   Fax: (42 7) 377 436, 375 454

3. Ministry of Health
   Limbova 2
   83101 BRATISLAVA

4. Slovak Agriculture and Food Inspection
   Priemyselna 4
   82409 BRATISLAVA

5. State Veterinary Administration
   Botanicka 17
   821 09 BRATISLAVA

6. Food Research Institute
   Priemyslena 4
   83101 BRATISLAVA

7. Ministry of Soil Management
   Dobrovicova 12
   81266 BRATISLAVA
   Tel: (42 7) 456 556
   Fax: (42 7) 365 110

8. Ecotoxicological Centre Bratislava Ltd.
   Attn: Mr M Murin
   Nadrazna 37
   900 28 IVANKA PRI DUNAJI
   Tel: (42 7) 943 712
   Fax: (42 7) 943 128, 945 223

9. Ms. Tatiana Plesnikova
   Infoterra NFP
   The Enviro-information Branch
   Ministry of Environment
   P.O. Box 1
   84002 BRATISLAVA
   Tel: (42 7) 786 691
   Fax: (42 7) 787 115

10. Institute of Preventive and Clinical Medicine
    Attn: Prof. L Rosival
    Limbova 14
    83301 BRATISLAVA

Slovenia
Republika Slovenija

Responsibility	
Main authority responsible for approval/registration of pesticides	
Main authority responsible for pesticide use directives, permits and training	
Main authority responsible for establishing national maximum pesticide residue limits	
Main authority responsible for control of pesticide residues in food	
Main laboratory for analysis of pesticide residues in food	
Main laboratory for quality control of pesticide products	
Main organisation responsible for applied research regarding pesticide use. Information on related regional research	
Designated national authority for notifications and information regarding banned and severely restricted pesticides including the PIC procedure	-
Codex Alimentarius contact point	2
National IRPTC correspondent	-
Infoterra national focal point	3
National IFCS focal point	1
National association of pesticide manufacturers	-

Note

Poisons information/control centre: University Medical Centre, Zaloska 7, 61000 Ljubljana, Tel: (38) 61-302-456 [Head: Dr. M. Mozina]

Responsible organisations

1. Ministry of Health
   Department for Dangerous
   and Toxic Substances
   Commission on Poisons
   Attn: Mr. Vojislav Suc
   Stefanova 5/II
   61000 LJUBLJANA
   Tel: (386 61) 1251028
   Fax: (386 61) 217752

2. Dr. Marusa Adamic
   Hygiene Specialist
   Institute for Public Health and
   Social Welfare
   Trubarjeva 2
   61000 LJUBLJANA
   Tel: (38 661) 1323245
   Fax: (38 661) 323955

3. Dr. Jure Besenicar
   Infoterra NFP
   Ministry for the Environment
   and Regional Planning
   Zupanciceva 6
   LJUBJANA
   Tel: (386 61) 214 827
   Fax: (386 61) 224 548

Solomon Islands

Responsibility	
Main authority responsible for approval/registration of pesticides	1
Main authority responsible for pesticide use directives, permits and training	
Main authority responsible for establishing national maximum pesticide residue limits	-
Main authority responsible for control of pesticide residues in food	2
Main laboratory for analysis of pesticide residues in food	-
Main laboratory for quality control of pesticide products	
Main organisation responsible for applied research regarding pesticide use. Information on related regional research	3
Designated national authority for notifications and information regarding banned and severely restricted pesticides including the PIC procedure	1
Codex Alimentarius contact point	-
National IRPTC correspondent	-
Infoterra national focal point	4
National IFCS focal point	-
National association of pesticide manufacturers	

Responsible organisations

1. The Registrar of Pesticides
   Director of Research
   Dodo Creek Research Station
   Ministry of Agriculture and Lands
   P.O. Box G13
   HONIARA
   Fax: 21955

2. Ministry of Agriculture and Lands
   P.O. Box G13
   HONIARA

3. Department of Research
   Ministry of Agriculture and Lands
   P.O. Box G13
   HONIARA

4. Infoterra NFP
   The Environment and Conservation Division
   Ministry of Forest, Environment and Conservation
   P.O. Box G24
   HONIARA
   Tel: (677) 25848
   Fax: (677) 21245

Somalia
Soomaaliya

Responsibility

Main authority responsible for approval/registration of pesticides	
Main authority responsible for pesticide use directives, permits and training	
Main authority responsible for establishing national maximum pesticide residue limits	
Main authority responsible for food control of pesticide residues in food	
Main laboratory for analysis of pesticide residues in food	
Main laboratory for quality control of pesticide products	
Main organisation responsible for applied research regarding pesticide use. Information on related regional research	
Designated national authority for notifications and information regarding banned and severely restricted pesticides including the PIC procedure	-
Codex Alimentarius contact point	-
National IRPTC correspondent	-
Infoterra national focal point	1
National IFCS focal point	-
National association of pesticide manufacturers	-

Responsible organisations

1. Mr. Ismail Deria Gammadid
   Infoterra NFP
   Director,
   Department of Forestry
   National Range Agency
   c/o UNDP
   P.O. Box 24
   MOGADISHU

South Africa
Republic of South Africa

Responsibility	
Main authority responsible for approval/registration of pesticides	1, 2
Main authority responsible for pesticide use directives, permits and training	
Main authority responsible for establishing national maximum pesticide residue limits	1, 12
Main authority responsible for control of pesticide residues in food	3, 4, 5, 6, 7
Main laboratory for analysis of pesticide residues in food	3, 4, 5, 6, 7, 8
Main laboratory for quality control of pesticide products	
Main organisation responsible for applied research regarding pesticide use. Information on related regional research	3
Designated national authority for notifications and information regarding banned and severely restricted pesticides including the PIC procedure	-
Codex Alimentarius contact point	9
National IRPTC correspondent	-
Infoterra national focal point	13
National IFCS focal point	10
National association of pesticide manufacturers	11

Note

Poisons information/control centres in Cape Town and Johannesburg

Responsible organisations

1. The Registrar for Pesticides
National Department of Agriculture
Private Bag X343
PRETORIA 0001
Tel: (27 12) 319 7000
Fax: (27 12) 319 7179

2. The Chairman
INDAC
Private Bag X343
PRETORIA 0001
Tel: (27 12) 319 7303
Fax: (27 12) 319 7179

3. Agricultural Research Council
Plant Protection Research Institute
Private Bag X134
PRETORIA 0001
Tel: (27 12) 808 0952
Fax: (27 12) 323 5275

4. Plant and Quality Control
Division of Agricultural Product Standards
Private Bag X116
PRETORIA 0001
Tel: (27 12) 319 6511
Fax: (27 12) 319 6055

5. South African Bureau of Standards
Private Bag X191
PRETORIA 0001
Tel: (27 12) 341 1311
Fax: (27 12) 344 1568

6. Health Chemistry Laboratories
P.O. Box 668
CAPE TOWN 8000
Tel: (27 21) 254083

7. City Health Department
Branch Laboratory and Technical Services
P.O. Box 1477
JOHANNESBURG 2000
Tel: (27 11) 728 7373

8. Working Group on Pesticide Analysis (WGPA)
Private Bag X134
PRETORIA 0001
Tel: (27 12) 808 0952
Fax: (27 12) 323 5275

9. Director,
Food and Chemicals
Department of Health
Private Bag X828
PRETORIA 0001
Tel: (27 12) 312 0554
Fax: (27 12) 325 8721

10. Director,
Pollution Control
Department of Environmental Affairs and Tourism
Private Bag X447
PRETORIA 0001
Tel: (27 12) 310 3911
Fax: (27 12) 322 6287

11. Agriculture and Veterinary Chemicals Association of South Africa (AVCASA)
P.O. Box 1995
HALFWAY HOUSE 1685
Tel: (27 11) 805 2000
Fax: (27 11) 805 2222

12. Department of Health
Private Bag X828
PRETORIA 0001
Tel: (27 12) 312 0000
Fax: (27 12) 325 5706

13. Dr. J.R. Pretorius
Infoterra NFP
Chief,
Environmental Management Directorate
Department of Environmental Affairs and Tourism
Private Bag X447
PRETORIA
Tel: (27 12) 310 3713
Fax: (27 12) 322 6287

Spain

Reino de España

Responsibility	
Main authority responsible for approval/registration of pesticides	1, 2
Main authority responsible for pesticide use directives, permits and training	1, 2
Main authority responsible for establishing national maximum pesticide residue limits	1, 2
Main authority responsible for control of pesticide residues in food	3, 4
Main laboratory for analysis of pesticide residues in food	3, 4
Main laboratory for quality control of pesticide products	5
Main organisation responsible for applied research regarding pesticide use. Information on related regional research	1, 7
Designated national authority for notifications and information regarding banned and severely restricted pesticides including the PIC procedure	8
Codex Alimentarius contact point	9
National IRPTC correspondent	10
Infoterra national focal point	11
National IFCS focal point	12
National association of pesticide manufacturers	6

Note

Poisons information/control centres in Barcelona, Madrid and Sevilla

Responsible organisations

1. Subdirección General de Sanidad Vegetal
   Ministerio de Agricultura, Pesca y Alimentación
   c/Velazquez, 147
   E-28002 MADRID
   Tel: (34 1) 347 8274
   Fax: (34 1) 347 8248

2. Subdireccion General de Sanidad Ambiental
   Ministerio de Sanidad y Consumo
   Paseo del Prado, 18
   E-28014 MADRID
   Tel: (34 1) 596 2084
   Fax: (34 1) 596 4409

3. Instituto de Salud Carlos III
   Ministerio de Sanidad y Consumo
   Ctra. de Majadahonda a Pozuelo, Km. 2
   E-28023 MAJADAHONDA (MADRID)
   Tel: (34 1) 638 2439
   Fax: (34 1) 638 0613

4. Laboratorio Agroalimentario del Estado
   Ministerio de Agricultura, Pesca y Alimentación
   Avda. del Padre Huidobro s/n
   Ctra. de la Coruña,
   Km. 10,700
   E-28071 MADRID
   Tel: (34 1) 347 4945

5. Centro de Investigacion y Control de la Calidad
   Instituto Nacional del Consumo
   Ministerio de Sanidad y Consumo
   Avda. de Cantabria, s/n
   28042 BARAJAS (MADRID)
   Tel: (34 1) 747 1500
   Fax: (34 1) 747 9517

6. Asociación Empresarial para la Protección de las Plantas (AEPLA)
   Almagro 44
   E-28010 MADRID
   Tel: (34 1) 310 0238,
   410 3449
   Fax: (34 1) 319 7817

7. Instituto Nacional de Investigaciones Agrarias
   Ministerio de Agricultura, Pesca y Alimentación
   c/José Abascal, 56
   E-28006 MADRID
   Tel: (34 1) 347 3900
   Telex: 48989 inia

8. Director General de Salud Publica
   Subdirección General de Sanidad Ambiental
   Ministerio de Sanidad y Consumo
   Paseo del Prado, 18
   E-28014 MADRID
   Tel: (34 1) 596 2011
   Fax: (34 1) 596 4409

9. Comision Interministerial para la Ordenacion Alimentaria (CIOA)
Ministerio de Sanidad y Consumo
Paseo del Prado, 18
28014 MADRID
Tel: (34 1) 596 1346
Fax: (34 1) 596 1762

10. Subdireccion General de Sanidad Ambiental
Attn. Dr E Tacoronte
Ministerio de Sanidad y Consumo
Paseo del Prado, 18
E-28014 MADRID
Tel: (34 1) 596 2011
Fax: (34 1) 596 4409

11. Ms Carmen Casal
CNC de Infoterra
Dirección General de Politica Ambiental
Ministerio de Obras Públicas, Transportes y Medio Ambiente Paseo de la Castellana, 67
E-28071 MADRID
Tel: (34 1) 597 7455
Fax: (34 1) 597 8511

12. Dr. F. Vargas Marcos
Subdirector General de Sanidad Ambiental
Paseo del Prado, 18
28014 MADRID
Tel: (34 1) 596 2084
Fax: (34 1) 596 44 09

Sri Lanka

Sri Lanka Prajathanthrika Samajavadi Janarajaya

Responsibility	
Main authority responsible for approval/registration of pesticides	1
Main authority responsible for pesticide use directives, permits and training	1, 9, 10, 11, 12, 13
Main authority responsible for establishing national maximum pesticide residue limits	1
Main authority responsible for control of pesticide residues in food	2
Main laboratory for analysis of pesticide residues in food	4, 5
Main laboratory for quality control of pesticide products	3
Main organisation responsible for applied research regarding pesticide use. Information on related regional research	9, 10, 11, 12, 13
Designated national authority for notifications and information regarding banned and severely restricted pesticides including the PIC procedure	1
Codex Alimentarius contact point	2
National IRPTC correspondent	5
Infoterra national focal point	6
National IFCS focal point	8
National association of pesticide manufacturers	7

Note

Poisons information/control centre: General Hospital, Colombo 8, Tel: (94) 168 6143, Fax: (94) 159 9231 [Head: Dr. R. Fernando]

Responsible organisations

1. The Registrar of Pesticides
   Pesticides Registration Office
   P.O. Box 49
   Getambe
   PERADENIYA
   Tel: (94 8) 88135
   Fax: (94 8) 88206

2. Director General of Health Services
   Food Advisory Committee
   Ministry of Health
   5th Floor
   Inland Revenue Building
   G.P.O. Box 513
   COLOMBO 10
   Tel: (94 1) 432 050
   Fax: (94 1) 440 399

3. Sri Lanka Standards Institution
   53 Dharmapala Mawatha
   COLOMBO 03
   Tel: (94 1) 26051

4. Ceylon Institute of Scientific and Industrial Research
   363 Bauddhaloka Mawatha
   COLOMBO 07
   Tel: (94 1) 93807

5. The Government Analyst
   Government Analyst's Department
   Torrington Square
   COLOMBO 07
   Tel: (94 1) 693 562
   Fax: (94 1) 692 309

6. Infoterra NFP
   Central Environmental Authority
   Parisara Mawatha
   Maligawatte New Town
   P.O. Box 2205
   COLOMBO 10
   Tel: (94 1) 449 455
   Fax: (94 1) 446 749

7. Pesticides Association of Sri Lanka (PASL)
   25 Foster Lane
   P.O. Box 70
   COLOMBO 10
   Tel: (94 1) 699 087
   Fax: (94 1) 697 202, 691 679

8. Central Environmental Authority
   Ministry of Environment and Parliamentary Affairs
   Maligawatte New Town
   COLOMBO 10
   Tel: (94 1) 583 290
   Fax: (94 1) 502 566

9. Department of Agriculture
   Old Galaha Road
   PERADENIYA
   Tel: (94 8) 88331
   Fax: (94 8) 88206

10. Tea Research Institute
    TALAWAKELLE
    Tel: (94 52) 8385, 8386
    Fax: (94 52) 8311, 8229

11. Coconut Research Institute
    Bandirippuwa Estate
    LUNUWILA
    Tel: (94 30) 3795
    Fax: (94 31) 7195

Responsible organisations

12. Rubber Rersearch Institute
"Dartonfield"
AGALAWATTA
Tel: (94 34) 71426

13. Department of Export Agriculture
1095 Kandy Road
Getambe
PERDENIYA
Tel: (94 08) 88364
Fax: (94 08) 88738

Sudan

Jumhuriyat as-Sudan, Republic of the Sudan

Responsibility	
Main authority responsible for approval/registration of pesticides	1
Main authority responsible for pesticide use directives, permits and training	1, 7
Main authority responsible for establishing national maximum pesticide residue limits	5
Main authority responsible for control of pesticide residues in food	2, 3, 4
Main laboratory for analysis of pesticide residues in food	2
Main laboratory for quality control of pesticide products	8
Main organisation responsible for applied research regarding pesticide use. Information on related regional research	7, 10
Designated national authority for notifications and information regarding banned and severely restricted pesticides including the PIC procedure	1
Codex Alimentarius contact point	4
National IRPTC correspondent	3
Infoterra national focal point	9
National IFCS focal point	-
National association of pesticide manufacturers	6

Responsible organisations

1. The Registrar
   National Pesticide Committee
   Plant Protection Directorate
   P.O. Box 14
   KHARTOUM NORTH
   Tel: 612 665, 612 666

2. Director,
   National Chemical Laboratories
   P.O. Box 287
   KHARTOUM
   Tel: 72991

3. The Head,
   Contaminants Department
   National Chemical Laboratories
   Ministry of Health
   P.O. Box 287
   KHARTOUM

4. The Secretary
   National Codex Committee
   Committee of Chemical Laboratories
   Ministry of Health and Welfare
   P.O. Box 287
   KHARTOUM

5. Ministry of Health
   P.O. Box 303
   KHARTOUM
   Tel: 73000

6. Sudan Agrochemical Association (SAGA)
   (address unknown)

7. Agricultural Research Corporation
   P.O. Box 126
   WAD MEDANI
   Tel: 2226

8. National Pesticides Laboratories
   P.O. Box 126
   WAD MEDANI

9. Ms. Cecile Wesley
   Director,
   Documentation and Information Centre
   National Centre for Research
   P.O. Box 2404
   KHARTOUM
   Tel: (249 11) 70776
   Fax: (249 11) 70701

10. Department of Crop Protection
    University of Khartoum
    Faculty of Agriculture
    P.O. Box 32
    SHAMBAT

Suriname
Republiek Suriname

Responsibility	
Main authority responsible for approval/registration of pesticides	1
Main authority responsible for pesticide use directives, permits and training	
Main authority responsible for establishing national maximum pesticide residue limits	1, 2
Main authority responsible for control of pesticide residues in food	3, 4
Main laboratory for analysis of pesticide residues in food	5
Main laboratory for quality control of pesticide products	
Main organisation responsible for applied research regarding pesticide use. Information on related regional research	6
Designated national authority for notifications and information regarding banned and severely restricted pesticides including the PIC procedure	7
Codex Alimentarius contact point	8
National IRPTC correspondent	9
Infoterra national focal point	-
National IFCS focal point	-
National association of pesticide manufacturers	-

Responsible organisations

1. Ministry of Agriculture, Animal Husbandry and Fisheries
   P.O. Box 160
   PARAMARIBO
   Tel: (597) 472 442
   Fax: (597) 478 986

2. Ministry of Health
   Gravenstraat 64
   P.O. Box 201
   PARAMARIBO
   Tel: (597) 410 441
   Fax: (597) 410 702

3. Keuringsdienst van Voedingmiddelen en Waren
   P.O. Box 1911
   PARAMARIBO
   Tel: (597) 499 061
   Fax: (597) 497 835

4. Veterinaire Inspectie
   Beekhuizen
   P.O. Box 1016
   PARAMARIBO
   Tel: (597) 479 329
   Fax: (597) 474 407

5. Centraal Laboratorium
   Afdeling Residue-Analyse
   P.O. Box 1911
   PARAMARIBO
   Tel: (597) 499 061

6. Ministry of Agriculture, Animal Husbandry and Fisheries
   Research Divison
   P.O. Box 160
   PARAMARIBO
   Tel: (597) 472 442
   Fax: (597) 478 986

7. Pesticides Division
   Ministry of Agriculture, Animal Husbandry and Fisheries
   P.O. Box 160
   PARAMARIBO
   Tel: (597) 472 442
   Fax: (597) 478 986

8. Ir. G. Hindoire
   Head, Division of Foreign Relations
   Ministry of Agriculture, Livestock and Fisheries
   Cultuurttuinlaan
   P.O. Box 1807
   PARAMARIBO
   Tel: (597) 478 986
   Fax: (597) 478 986

9. The Chairman
   Pesticide Working Group
   Central Laboratory
   P.O. Box 1911
   PARAMARIBO

Swaziland
Umbuso weSwatini, Kingdom of Swaziland

Responsibility	
Main authority responsible for approval/registration of pesticides	-
Main authority responsible for pesticide use directives, permits and training	
Main authority responsible for establishing national maximum pesticide residue limits	-
Main authority responsible for control of pesticide residues in food	-
Main laboratory for analysis of pesticide residues in food	-
Main laboratory for quality control of pesticide products	-
Main organisation responsible for applied research regarding pesticide use. Information on related regional research	1
Designated national authority for notifications and information regarding banned and severely restricted pesticides including the PIC procedure	-
Codex Alimentarius contact point	2
National IRPTC correspondent	3
Infoterra national focal point	4
National IFCS focal point	-
National association of pesticide manufacturers	-

Responsible organisations

1. Ministry of Agriculture and Cooperatives
   Agricultural Research Division
   P.O. Box 4
   MALKERNS
   Tel: 83017, 83038
   Fax: 83360

2. The Principal Secretary
   Att: Director of Health Services
   Ministry of Health
   P.O. Box 5
   MBABANE
   Tel: (0268)42431
   Fax: (0268)42092

3. The Factories Inspector
   Department of Labour
   P.O. Box 198
   MBABANE
   Tel: 41971
   Fax: 41966

4. Mr J.D. Vilakati
   Infoterra NFP
   Executive Director
   Environmental Authority
   Ministry of Natural Resources and Energy
   P.O. Box 57
   MBABANE
   Tel: (268) 46244
   Fax: (268) 42436, 41719

Sweden
Sverige

Responsibility

Main authority responsible for approval/registration of pesticides	1
Main authority responsible for pesticide use directives, permits and training	2, 3, 4
Main authority responsible for establishing national maximum pesticide residue limits	5
Main authority responsible for control of pesticide residues in food	5
Main laboratory for analysis of pesticide residues in food	5
Main laboratory for quality control of pesticide products	6
Main organisation responsible for applied research regarding pesticide use. Information on related regional research	7
Designated national authority for notifications and information regarding banned and severely restricted pesticides including the PIC procedure	1
Codex Alimentarius contact point	5
National IRPTC correspondent	1
Infoterra national focal point	8
National IFCS focal point	9
National association of pesticide manufacturers	10

Note

Poisons information/control centre: Giftinformationscentralen, S-17176 Stockholm, Tel: +46-8-6100511, Fax: +46-8-327584 [Head: Dr. H Persson]

Responsible organisations

1. The National Chemicals
   Inspectorate - KemI
   P.O. Box 1384
   S-171 27 SOLNA
   Tel: (46 8) 730 5700
   Fax: (46 8) 735 7698

2. Swedish Board of Agriculture
   S-55182 JÖNKÖPING
   Tel: (46 36) 155 000
   Fax: (46 36) 190 546

3. Swedish Environmental
   Protection Agency
   S-10648 STOCKHOLM
   Tel: (46 8) 698 1000
   Fax: (46 8) 202 925

4. National Board of
   Occupational Safety and
   Health
   S-17184 SOLNA
   Tel: (46 8) 730 9000
   Fax: (46 8) 730 1967

5. National Food Administration
   P.O. Box 622
   S-751 26 UPPSALA
   Tel: (46 18) 175 500
   Fax: (46 18) 105 848

6. Organic Environmental
   Chemistry Section
   Department of Environmental
   Assessment
   Swedish University of
   Agricultural Sciences
   P.O. Box 7050
   S-75007 UPPSALA
   Tel: (46 18) 673 030
   Fax: (46 18) 302 753

7. Swedish University of
   Agricultural Sciences
   P.O. Box 7070
   S-750 07 UPPSALA
   Tel: (46 18) 671 000
   Fax: (46 18) 673 509

8. Infoterra/Sweden
   Attn: Dr. A. Sjöqvist
   Swedish Environmental
   Protection Agency
   Blekholmsterassen 36
   S-106 48 STOCKHOLM
   Tel: (46 8) 698 1000
   Fax: (46 8) 698 1400

9. The Director-General
   The National Chemicals
   Inspectorate - KemI
   P.O. Box 1384
   S-17127 SOLNA
   Tel: (46 8) 730 5700
   Fax: (46 8) 735 7698

10. The Association of Swedish
    Plant and Wood Protection
    Industries - IVT
    P.O. Box 5501
    S-11485 STOCKHOLM
    Tel: (46 8) 783 8000
    Fax: (46 8) 663 6323

Switzerland

Confédération Suisse, Schweizerische Eidgenossenschaft, Confederazione Svizzera

Responsibility	
Main authority responsible for approval/registration of pesticides	1
Main authority responsible for pesticide use directives, permits and training	1, 3
Main authority responsible for establishing national maximum pesticide residue limits	1, 2, 3
Main authority responsible for control of pesticide residues in food	3
Main laboratory for analysis of pesticide residues in food	3
Main laboratory for quality control of pesticide products	1
Main organisation responsible for applied research regarding pesticide use. Information on related regional research	1, 4
Designated national authority for notifications and information regarding banned and severely restricted pesticides including the PIC procedure	5
Codex Alimentarius contact point	6
National IRPTC correspondent	7
Infoterra national focal point	5
National IFCS focal point	8
National association of pesticide manufacturers	9

Note

Poisons information/control centre: Centre Suisse d'Information Toxicologique, Klosbachstrasse 107, 8030 Zurich, Tel: (41) 1 251 6666, Fax: (41) 1 252 8833, [Head: Dr P J Meier-Abt]

Responsible organisations

1. Eidg. Forschungsanstalt für Obst-, Wein- und Gartenbau/
Station de Recherches en Arboriculture, Viticulture et Horticulture
CH-8820 WÄDENSWIL
Tel: (41 1) 783 6111
Fax: (41 1) 780 6341

2. Bundesamt für Gesundheitswesen/
Office Federale de la Sante Publique
Fachstelle Toxicologie/
Service Toxicologie
Institut für Veterinär-pharmakologie und Toxicologie
Winterthurerstrasse 260
CH-8057 ZÜRICH
Tel: (41 1) 257 6105
Fax: (41 1) 257 6107

3. Bundesamt für Gesundheitswesen/
Office Fédéral de la Santé Publique
Abteilung Lebensmittel-wissenschaft/
Division Sciences des Alimentaires
Haslerstrasse 16
CH-3008 BERNE
Tel: (41 31) 322 9586 (for general information)
Tel: (41 31) 322 9552 (for tolerances)
Fax: (41 31) 322 9574

4. Landwirtschaftliche Forschungsanstalt Changins/
Station de Recherches Agronomiques de Changins
CH-1260 NYON
Tel: (41 22) 363 4444
Fax: (41 22) 362 1325

5. Service des affaires internationales
Office féderal de l'environnement, des forets et du paysage (OFEFP)
Hallwylstrasse 4
3003 BERNE
Tel: (41 31) 322 9973
Fax: (41 31) 322 9981

6. Secrétariat
Comité National Suisse du Codex Alimentarius
Office Fédéral de la Santé Publique
Haslerstrasse 16
CH-3008 BERNE
Tel: (41 31) 322 9572
Fax: (41 31) 322 9574

7. Bundesamt fur Umwelt, Wald und Landschaft/
Office Fédéral de l'Environnement, des Forêts et du Paysage
Abteilung Stoffe und Bodenschutz/
Division Substances et Protection du Sol
CH-3003 BERNE
Tel: (41 31) 322 9349
Fax: (41 31) 324 7978

Responsible organisations

8. Mr G Karlaganis
Head, Division of Substances and Soil Protection
Federal Office of Environment, Forests and Landscape
CH-3003 BERNE
Tel: (41-31) 322 6955
Fax: (41-31) 324 7978

9. Schweizerische Gesellschaft für Chemische Industrie/
Société Suisse des Industries Chemiques (SSIC)
Nordstrasse 15
Postfach
CH-8035 ZÜRICH
Tel: (41 1) 363 1030, 368 1711
Fax: (41 1) 363 1018, 368 1770

Syria

al-Jumhuriyah al-Arabiyah as-Suriyah, Syrian Arab Republic

Responsibility	
Main authority responsible for approval/registration of pesticides	1, 2
Main authority responsible for pesticide use directives, permits and training	
Main authority responsible for establishing national maximum pesticide residue limits	3
Main authority responsible for control of pesticide residues in food	1, 4
Main laboratory for analysis of pesticide residues in food	1, 4
Main laboratory for quality control of pesticide products	
Main organisation responsible for applied research regarding pesticide use. Information on related regional research	5, 6
Designated national authority for notifications and information regarding banned and severely restricted pesticides including the PIC procedure	7
Codex Alimentarius contact point	3
National IRPTC correspondent	8
Infoterra national focal point	9
National IFCS focal point	10
National association of pesticide manufacturers	-

Responsible organisations

1. Ministry of Health
   DAMASCUS

2. Ministry of Agriculture
   DAMASCUS

3. Syrian Arab Organization for Standardization and Metrology
   P.O. Box 11836
   DAMASCUS

4. Ministry of Supply
   DAMASCUS

5. Central Analytical Laboratory
   Ministry of Industry
   DAMASCUS

6. Faculty of Pharmacy
   University of Damascus
   DAMASCUS

7. The Co-ordinator
   Ministry of State for Environmental Affairs
   Council of the Prime Minister
   Shahbandar Street
   DAMASCUS
   Tel: (963 11) 662 3148
   Fax: (963 11) 662 2309

8. Mr. Fouad El-Ok
   Ministry of the Environment
   Environmental and Scientific Research Centre
   P.O. Box 38782
   Mazzeh-Western Villas
   DAMASCUS

9. Ms Lama Abdul Salam
   Infoterra National
   Focal Point
   Ministry of State for the Environment
   P.O. Box 3773
   DAMASCUS
   Tel: (963 11) 330 510
   Fax: (963 11) 335 645

10. Mr. Fouad El-Ok
    General Commission for Environmental Affairs
    Tolyani Street
    P.O. Box 3773
    DAMASCUS
    Tel: (963 11) 330 510
    Fax: (963 11) 335 645

Taiwan
Chung-hua Min-kuo

Responsibility

Main authority responsible for approval/registration of pesticides	1
Main authority responsible for pesticide use directives, permits and training	1, 3
Main authority responsible for establishing national maximum pesticide residue limits	1, 2
Main authority responsible for control of pesticide residues in food	2
Main laboratory for analysis of pesticide residues in food	4, 6
Main laboratory for quality control of pesticide products	7
Main organisation responsible for applied research regarding pesticide use. Information on related regional research	3
Designated national authority for notifications and information regarding banned and severely restricted pesticides including the PIC procedure	-
Codex Alimentarius contact point	1, 2
National IRPTC correspondent	1
Infoterra national focal point	-
National IFCS focal point	-
National association of pesticide manufacturers	5, 8

Note

Poisons information/control centre: Division of Clinical Toxicology, Department of Medicine, Veterans General Hospital, Shih-Pai Road 201, sec.2, Taipei 11217, Tel: (886) 2875 7525, Fax: (886) 2873 9193 [Head: Dr. J.F.Deng]

Responsible organisations

1. Council of Agriculture
   Executive Yuan (COA)
   37 Nanhai Road
   10728 TAIPEI
   Tel: (02) 381 2991
   Fax: (02) 311 7621

2. Bureau of Food Sanitation
   Department of Health
   100 Aikuo East Road
   TAIPEI
   Tel: (02) 396 5625
   Fax: (02) 392 9723

3. Provincial Department of Agriculture and Forestry (PDAF)
   8 Kuanghua Road
   Chunghsing Village
   NANTOU HSIEN
   Tel: (049) 332 380
   Fax: (049) 351 011

4. Taiwan Agricultural Chemicals and Toxic Substances Research Institute (TACTRI)
   11 Kuang Ming Road
   Wufeng
   413 TAICHUNG HSIEN
   Tel: (886 4) 330 2101
   Fax: (886 4) 332 3073

5. Taiwan Agrochemical Industry Association (TAIA)
   7th Floor, No 96, Sec 2
   Chien Kuo N Road
   TAIPEI
   Tel: (886 2) 5034666
   Fax: (886 2) 5034405

6. National Laboratory of Foods and Drugs
   161 Quenyang Street
   Nankang
   TAIPEI
   Tel: (02) 785 8283
   Fax: (02) 785 8608

7. Bureau of Commodity and Quarantine (BCIQ)
   4 Chinan Road. Sec.1
   TAIPEI
   Tel: (02) 343 1700
   Fax: (02) 393 2324

8. Taipai Agricultural Chemicals Association
   109 TAIPEI
   Fax: (886 2) 537 3589

Tajikistan
Jumhurii Tojikistan

Responsibility	
Main authority responsible for approval/registration of pesticides	
Main authority responsible for pesticide use directives, permits and training	
Main authority responsible for establishing national maximum pesticide residue limits	
Main authority responsible for control of pesticide residues in food	
Main laboratory for analysis of pesticide residues in food	
Main laboratory for quality control of pesticide products	
Main organisation responsible for applied research regarding pesticide use. Information on related regional research	
Designated national authority for notifications and information regarding banned and severely restricted pesticides including the PIC procedure	1
Codex Alimentarius contact point	-
National IRPTC correspondent	1
Infoterra national focal point	2
National IFCS focal point	-
National association of pesticide manufacturers.	-

Responsible organisations

1. Chairman,
   Khatlon Regional Committee of Nature Protection
   Ministry of Nature Protection
   Bokhtar 12
   DUSHANBE
   Tel: (7 377) 232 878
   Fax: (7 377) 211 839

2. Infoterra NFP
   Ministry of Environment Protection
   Ordzonikidze Str., 12
   DUSHANBE
   Tel: (377 2) 232 878
   Fax: (377 2) 275 581

Tanzania

Jamhuri ya Muungano wa Tanzania, United Republic of Tanzania

Responsibility	
Main authority responsible for approval/registration of pesticides	1
Main authority responsible for pesticide use directives, permits and training	1
Main authority responsible for establishing national maximum pesticide residue limits	1, 3
Main authority responsible for control of pesticide residues in food.	1, 3, 13
Main laboratory for analysis of pesticide residues in food	1, 3, 4
Main laboratory for quality control of pesticide products	1
Main organisation responsible for applied research regarding pesticide use. Information on related regional research	1, 2, 6, 7, 8, 9, 10, 11, 12
Designated national authority for notifications and information regarding banned and severely restricted pesticides including the PIC procedure	1
Codex Alimentarius contact point	3
National IRPTC correspondent	4
Infoterra national focal point	15
National IFCS focal point	5
National association of pesticide manufacturers	14

Note

Poisons information/control centre: Tanzania Drug and Toxicology Information Service (TADATIS), Muhimbili Hospital, P.O. Box 65088, Dar-es-Salaam, Tel:(255 51) 26 211, Fax:(255 51) 46 163 [Head: Mr C Y S Mwasha]

Responsibility organisations

1. The Registrar of Pesticides
   Tropical Pesticides Research
   Institute (TPRI)
   P.O. Box 3024
   ARUSHA
   Tel: (057) 8813
   Fax: (057) 8217

2. Ministry of Agriculture
   P.O. Box 9192
   DAR-ES-SALAAM

3. Tanzania Bureau of Standards
   P.O. Box 9524
   DAR-ES-SALAAM
   Tel: 49041
   Fax: 48051

4. Chief Government Chemist
   Government Chemical
   Laboratory
   Ministry of Health and Social
   Welfare
   P.O. Box 164
   DAR-ES-SALAAM
   Tel: 25051

5. The Factories Inspectorate
   Ministry of Labour and Youth
   Development
   P.O. Box 9014
   DAR-ES-SALAAM
   Tel: (255-51) 29973
   Fax: (255-51) 66004

6. Ukiriguru Research Station
   P.O Box 1433
   MWANZA

7. Uyole Agricultural Centre
   P.O Box 400
   MBEYA

8. Tumbi Research Station
   P.O. Box 306
   TABORA

9. Sokoine University of
   Agriculture
   Faculty of Agriculture
   P.O. Box 3000
   MOROGORO

10. Maruku Research Station
    P.O. Box 127
    BUKOBA

11. Lyamungu Research Station
    P.O. Box 3004
    MOSHI

12. Ilonga Research Station
    Private Bag
    KILOSA

13. National Food Control
    Commission
    Ministry of Health
    DAR-ES-SALAAM

14. Agrochemical Association of
    Tanzania (AAT)
    P.O. Box 22141
    DAR-ES-SALAAM
    Tel: 22404, 31573, 38868

15. Mr G L Kamukala
    Director-General,
    National Environment
    Management Council
    Infoterra NFP
    P.O. Box 63154
    DAR-ES-SALAAM
    Tel: (051) 34603, 27815
    Fax: (051) 34603

Thailand
Muang Thai, Prathet Thai

Responsibility	
Main authority responsible for approval/registration of pesticides	1, 2, 9
Main authority responsible for pesticide use directives, permits and training	2, 5, 8
Main authority responsible for establishing national maximum pesticide residue limits	1, 3, 4
Main authority responsible for control of pesticide residues in food	1, 3
Main laboratory for analysis of pesticide residues in food	2, 3
Main laboratory for quality control of pesticide products	2, 3
Main organisation responsible for applied research regarding pesticide use. Information on related regional research	2, 3, 8
Designated national authority for notifications and information regarding banned and severely restricted pesticides including the PIC procedure	4, 10
Codex Alimentarius contact point	6
National IRPTC correspondent	4
Infoterra national focal point	8
National IFCS focal point	1
National association of pesticide manufacturers	7

Note

Poisons information/control centre: Department of Medicine, Ramathibodi Hospital, Rama VI Road, Bangkok 10400, Tel: (662) 245 5707, Fax: (662) 246 2123 [Head: Dr S Kaojarern]

Responsible organisations

1. The Food and Drug Administration
   Ministry of Public Health
   Tiwanon Road
   NONTHABURI 11000
   Tel: (66 2) 591 8458
   Fax: (66 2) 591 8452

2. Department of Agriculture
   Ministry of Agriculture and Cooperatives
   Phahon Yothin Road,
   Chatuchak
   BANGKOK 10900
   Tel: (66 2) 579 0151
   Fax: (66 2) 561 5024

3. Department of Medical Sciences
   Ministry of Public Health
   88/7 Soi Bamrasnadura Hospital
   Amphoe Muang
   NONTHABURI 11000
   Tel: (66 2) 589 9850,
   589 0022
   Fax: (66 2) 591 5449

4. Department of Pollution Control
   404 Phahon Yothin Center Building
   Phahon Yothin Road,
   Phayathai
   BANGKOK 10400
   Tel: (66 2) 619 2299
   Fax: (66 2) 619 2285

5. Department of Agriculture Extension
   Ministry of Agriculture and Cooperatives
   Phahon Yothin Road, Chatuchak
   BANGKOK 10900
   Tel: (66 2) 579 0121
   Fax: (66 2) 579 9540

6. Office of the National Codex Alimentarius Committee
   Ministry of Industry
   Rama VI Road, Ratchathewi
   BANGKOK 10400
   Tel: (66 2) 245 6181,
   202 3435
   Fax: (66 2) 246 4327

7. Thai Crop Protection Association
   Rasa Building, 14 Floor, Room 1405
   555 Phahon Yothin Road,
   Chatuchak
   BANKOK 10900
   Tel: (66 2) 937 0487
   Fax: (66 2) 937 0491

8. Department of Environmental Quality Promotion
   Ministry of Science, Technology and Environment
   60/1 Soi Phibun Watthana 7,
   Rama VI Road
   BANGKOK 10400
   Tel: (66 2) 277 7180
   Fax: (66 2) 279 0672,
   278 2693

Responsible organisations

9. Department of Fisheries
Ministry of Agriculture and Cooperatives
Phahon Yothin Road, Chatuchak
BANGKOK 10900
Tel: (66 2) 562 0600
Fax: (66 2) 562 0564

10. The Director-General,
Department of Agriculture
Ministry of Agriculture and Co-operatives
Rajadamnern Avenue
BANGKOK 10200
Tel: (66 2) 579 0586,
579 9636

Togo
République Togolaise

Responsibility	
Main authority responsible for approval/registration of pesticides	1
Main authority responsible for pesticide use directives, permits and training	
Main authority responsible for establishing national maximum pesticide residue limits	-
Main authority responsible for control of pesticide residues in food	2
Main laboratory for analysis of pesticide residues in food	-
Main laboratory for quality control of pesticide products	
Main organisation responsible for applied research regarding pesticide use. Information on related regional research	1, 3
Designated national authority for notifications and information regarding banned and severely restricted pesticides including the PIC procedure	4
Codex Alimentarius contact point	5
National IRPTC correspondent	4
Infoterra national focal point	6
National IFCS focal point	-
National association of pesticide manufacturers	-

Responsible organisations

1. Direction de la Protection des Végétaux
Cacaveli
B.P. 1263
LOMÉ
Tel: 213 773

2. Direction de la Nutrition et de Technologie Alimentaire
Cacaveli
B.P. 1242
LOMÉ
Tel: 214 118

3. Institute des Recherches Agronomiques Tropicales
Route de Kpalimé
B.P. 1163
LOMÉ
Tel: 215 287

4. Monsieur le Directeur
Direction de la Protection des Végétaux
Ministère de l'Aménagement Rural
B.P. 1263
LOMÉ
Tel: (228) 253 773, 257 537
Fax: (228) 211 008

5. M. le Chargé de Liaison du Codex Alimentarius
Division de la Nutrition et de la Technologie Alimentaire
B.P. 1242
LOMÉ

6. M. Folly Koudanou
PFN d'Infoterra
Ministère de l'Environnement et du Tourisme
B.P. 3114
LOMÉ
Tel: (228) 215 285, 215 352
Fax: (228) 218 927

Tonga
Kingdom of Tonga,
Puléanga Fakatu'i'o Tonga

Responsibility	
Main authority responsible for approval/registration of pesticides	1
Main authority responsible for pesticide use directives, permits and training	
Main authority responsible for establishing national maximum pesticide residue limits	-
Main authority responsible for control of pesticide residues in food	2
Main laboratory for analysis of pesticide residues in food	-
Main laboratory for quality control of pesticide products	
Main organisation responsible for applied research regarding pesticide use. Information on related regional research	1
Designated national authority for notifications and information regarding banned and severely restricted pesticides including the PIC procedure	3
Codex Alimentarius contact point	-
National IRPTC correspondent	1
Infoterra national focal point	5, 6
National IFCS focal point	4
National association of pesticide manufacturers	-

Responsible organisations

1. The Director
Ministry of Agriculture and Forestry
P.O. Box 14
NUKU'ALOFA
Tel: (676) 23402
Fax: (676) 24271

2. Director of Health
Ministry of Health
NUKU'ALOFA

3. Deputy Director and Head, Research Division
Ministry of Agriculture and Forestry
P.O. Box 14
NUKU'ALOFA
Tel: (676) 23038
Fax: (676) 24271

4. Dr. M Kinahoi
Chief Medical Officer
Ministry of Public Health
P.O. Box 59
NUKU'ALOFA
Tel: (676) 23200
Fax. (676) 24291

5. Infoterra NFP
The Secretary of Lands, Survey and Natural Resources
Ministry of Lands, Survey and Natural Resources
NUKU'ALOFA
Tel: (676) 23611
Fax: (676) 23216

6. Chief Secretary and Secretary to the Cabinet
Attn: Mr. Inoke F. Faletau
Prime Minister's Office
NUKU'ALOFA
Tel: (676) 21300
Fax: (676) 23888

Trinidad and Tobago

Republic of Trinidad and Tobago

Responsibility	
Main authority responsible for approval/registration of pesticides	1
Main authority responsible for pesticide use directives, permits and training	-
Main authority responsible for establishing national maximum pesticide residue limits	2
Main authority responsible for control of pesticide residues in food	2, 3
Main laboratory for analysis of pesticide residues in food	2
Main laboratory for quality control of pesticide products	2
Main organisation responsible for applied research regarding pesticide use. Information on related regional research	4, 5
Designated national authority for notifications and information regarding banned and severely restricted pesticides including the PIC procedure	6
Codex Alimentarius contact point	2
National IRPTC correspondent	6
Infoterra national focal point	-
National IFCS focal point	7
National association of pesticide manufacturers	-

Responsible organisations

1. Pesticides and Toxic
   Chemicals Control Board
   c/o Chief Chemist/Director of
   Food and Drugs
   115 Frederick Street
   PORT-OF-SPAIN
   Tel: 623 2476

2. The Chief Chemist
   Director of Food and Drugs
   Chemistry
   Food and Drugs Division
   Ministry of Health
   115 Frederick Street
   PORT-OF-SPAIN
   Tel: 623 2476

3. Chief Medical Officer
   Ministry of Health
   IDC Building
   Independence Square
   PORT-OF SPAIN
   Tel: 627 0010

4. University of the West Indies
   ST. AUGUSTINE
   Tel: 662 7171

5. Chief Technical Officer
   Ministry of Agriculture, Lands
   and Food Production
   St. Clair
   PORT-OF-SPAIN
   Tel: 622 1221

6. The Registrar
   Pesticides and Toxic Chemicals
   Chemistry/Food and Drugs
   Division
   Ministry of Health
   115 Frederick Street
   PORT-OF-SPAIN

7. Mr. Devnath Roopnarine
   Industrial Safety Officer
   Ministry of Labour and
   Cooperatives
   Riverside Plaza
   Besson Street
   PORT-OF-SPAIN

Tunisia
al-Jumhuriyah at-Tunisiyah

Responsibility

Main authority responsible for approval/registration of pesticides	1
Main authority responsible for pesticide use directives, permits and training	
Main authority responsible for establishing national maximum pesticide residue limits	1, 2
Main authority responsible for control of pesticide residues in food	3, 4, 5
Main laboratory for analysis of pesticide residues in food	-
Main laboratory for quality control of pesticide products	
Main organisation responsible for applied research regarding pesticide use. Information on related regional research	1
Designated national authority for notifications and information regarding banned and severely restricted pesticides including the PIC procedure	6
Codex Alimentarius contact point	7
National IRPTC correspondent	8
Infoterra national focal point	9
National IFCS focal point	10
National association of pesticide manufacturers	-

Note

Poisons information/control centre: Centre anti-Poisons, Centre d'Aide Medicale Urgente, Montfleury, Tunis, Tel: (216) 134 1655, Fax: (216) 124 0575 [Head: Dr M Yacoub]

Responsible organisations

1. Ministère de l'Agriculture
   Direction Générale de la Production Agricole
   Laboratoire de contrôles et d'Analyses des Pesticides
   30, Rue Alain Savary
   TN-1002 TUNIS
   Tel: (216 1) 287 133
   Fax: (216 1) 780 246

2. Institut National de la Normalisation et de la Propriété Industrielle (INNORPI)
   B.P.23
   TN-1002 TUNIS-BELVEDERE
   Tel: (216-1)781563

3. Laboratoire Général d'Analyses et d'Essais
   87 Av. Taieb El Mhiri
   TN-1002 TUNIS-BELVEDERE
   Tel: (216 1) 286 912, 785 285
   Fax: (216 1) 783 216

4. Institut National de Nutrition et de Technologie Alimentaire
   11 rue Djebel Lakhdar
   Bab Saadoun
   TN-1007 TUNIS
   Tel: (216 1) 570 665
   Fax: (216 1) 570 725

5. Laboratoire de Contrôles et d'Analyses des Pesticides
   30, Rue Alain Savary
   TN-1002 TUNIS - LE BELVEDERE
   Tel: (216 1) 681 639, 681 040

6. Monsieur le Directeur
   Agence nationale de protection de l'environnement (ANPE)
   12, Avenue Khéreddine Pacha
   TN-1002 TUNIS - LE BELVEDERE
   Tel: (216 1) 798 122, 799 201
   Fax: (216 1) 789 844

7. Dr A Jilani
   Directeur Générale Adjoint
   Institut National de la Normalisation et de la Propreté Industrielle (INNORPI)
   B.P. 23
   TN-1012 TUNIS - BELVEDERE
   Tel: (216 1) 891 742
   Fax: (216 1) 781 563

8. M. le Directeur
   Ministère de l'Economie Nationale
   Direction Genérale de l'Industrie
   Place du Gouvernement
   La Kasbah
   TN-1002 TUNIS
   Tel: (216 1) 892 313
   Fax: (216 1) 285 947

9. M. Nejib Daouas
   PFN d'Infoterra
   Ministère de l'Environnement et de l'Aménagement du Territoire
   Centre Urbain Nord,
   Imm I.C.F.
   TN-2080 ARIANA-TUNIS
   Tel: (216 1) 704 000
   Fax: (216 1) 702 431

10. Dr.Ing. N. Ben Mansour Chef,
Direction Générale de l'Industrie
Ministere de l'Economie Nationale
87 Av. Taïeb El Mhiri
TN-1002 TUNIS-BELVEDERE
Tel: (216 1) 892 313, 802 849
Fax: (216 1) 285 947

Turkey
Türkiye Cumhuriyeti

Responsibility	
Main authority responsible for approval/registration of pesticides	1
Main authority responsible for pesticide use directives, permits and training	1
Main authority responsible for establishing national maximum pesticide residue limits	1, 2
Main authority responsible for control of pesticide residues in food	1, 2
Main laboratory for analysis of pesticide residues in food	1, 2
Main laboratory for quality control of pesticide products	1
Main organisation responsible for applied research regarding pesticide use. Information on related regional research	1
Designated national authority for notifications and information regarding banned and severely restricted pesticides including the PIC procedure	7, 8
Codex Alimentarius contact point	1
National IRPTC correspondent	3
Infoterra national focal point	5
National IFCS focal point	9
National association of pesticide manufacturers	6

Note

Poisons information/control centre: Refik Saydam Hygiene Centre, Cemal Gursel Cad No. 18, 06100 Sihhiye, Ankara, Tel: (90 4) 433 7001, Fax: (90 4) 433 7000 [Head: Ms. N. Besbelli]

Responsible organisations

1. Ministry of Agriculture and Rural Affairs, General Directorate of Protection and Control
   Akay Cad. No: 3
   Bakanliklar
   ANKARA
   Tel: (312) 418 9835, 118 9835
   Fax: (312) 418 8005, 118 8005

2. Ministry of Health, Directory of the Poison Research
   Cemal Gürsel Cad. No: 18
   Cebeci
   ANKARA
   Tel: (312) 433 7001
   Fax: (312) 433 7000

3. The Director,
   Refik Saydam Hygiene Center
   Research Directorate of Poison Control
   General Directorate of Environment
   Ataturk Bulvari 143
   Bakanliklar
   ANKARA

4. Ministry of Health
   General Directorate of Public Health Services
   Sihhiye
   ANKARA
   Tel: (312) 435 5680
   Fax: (312) 434 4449

5. Ministry of Environment
   Foreign Relations Department
   Eskiesehir Yolu 8 Km
   06100 ANKARA
   Tel: (90 312) 285 1705
   Fax: (90 312) 285 3739

6. Society of Industrialists, Importers and Representatives for Pesticides - TISIT
   Defterdar Yokuu, Süngü Sokak
   Güneydou hani, Kat 3/15
   Tophane
   80040 ISTANBUL
   Tel: (901) 249 1748
   fax: (901) 251 4161

7. General Directorate of Environmental Pollution, Prevention and Control
   Ministry of Environment
   Eskisehir yolu 8 km.
   ANKARA
   Tel: (90 312) 285 0941
   Fax: (90 312) 285 5875

8. The Director,
   Research (Toxicology) Department
   Refik Saydam Hygiene Institute
   Cemal Gursel Cad. No.18
   Sihhiye
   06100 ANKARA
   Tel: (90 312) 433 7001
   Fax: (90 312) 433 7000

9. Dr. Nida Besbelli
   Head,
   Poison Research Center
   Refik Saydam Center
   Refik Saydam Health Institute
   ANKARA
   Tel: (90 312) 433 7001
   Fax: (90 312) 433 7000

Turkmenistan

Türkmenistan Jumhuriyäti

Responsibility

Main authority responsible for approval/registration of pesticides	
Main authority responsible for pesticide use directives, permits and training	
Main authority responsible for establishing national maximum pesticide residue limits	
Main authority responsible for control of pesticide residues in food	
Main laboratory for analysis of pesticide residues in food	
Main laboratory for quality control of pesticide products	
Main organisation responsible for applied research regarding pesticide use. Information on related regional research	
Designated national authority for notifications and information regarding banned and severely restricted pesticides including the PIC procedure	-
Codex Alimentarius contact point	-
National IRPTC correspondent	-
Infoterra national focal point	1
National IFCS focal point	-
National association of pesticide manufacturers	-

Responsible organisations

1. Mr. Nuri Babaevich Babaev
   Infoterra NFP
   Turkmenian Institute of
   Scientific and Technical
   Information
   (TURKMENNIINTI)
   Borodinskaya Str., 2
   ASHKHABAD 744000
   Tel: (363 2) 29 51 23
   Fax: (363 2) 29 59 43

Uganda
Republic of Uganda

Responsibility	
Main authority responsible for approval/registration of pesticides	1
Main authority responsible for pesticide use directives, permits and training	
Main authority responsible for establishing national maximum pesticide residue limits	1
Main authority responsible for control of pesticide residues in food	1,2
Main laboratory for analysis of pesticide residues in food	1
Main laboratory for quality control of pesticide products	
Main organisation responsible for applied research regarding pesticide use. Information on related regional research	3, 4
Designated national authority for notifications and information regarding banned and severely restricted pesticides including the PIC procedure	5
Codex Alimentarius contact point	6
National IRPTC correspondent	1
Infoterra national focal point	7
National IFCS focal point	8
National association of pesticide manufacturers	-

Responsible organisations

1. The Chief Government Chemist
   Laboratory of the Government Chemist
   Attn: Mr K E Okelo
   P.O. Box 2174
   KAMPALA
   Tel: 43303, 43222

2. Uganda National Food and Nutritional Council
   Ministry of Agriculture and Forestry
   P.O Box 102
   ENTEBBE
   Tel: (42) 20981

3. Commissioner for Tsetse Control
   Tsetse Control Departement
   Ministry of Animal Husbandry,
   Game and Fisheries
   P.O. Box 7003
   KAMPALA
   Tel: 33789, 57351

4. Kawanda Agricultural Research Station
   P.O Box 7065
   KAMPALA
   Tel: 67621

5. The Commissioner for Agriculture
   P.O Box 2
   ENTEBBE
   Tel: (256 42) 20981
   Fax: (256 42) 21047

6. The Executive Director
   Uganda National Bureau of Standards
   P.O. Box 6329
   KAMPALA

7. Ms. Elizabeth Nalubega Gowa
   Infoterra NFP
   National Environment Information Centre
   Ministry of Natural Resources
   P.O. Box 9629
   KAMPALA
   Tel: (256 41) 236817, 243613
   Fax: (256 41) 232680, 236819

8. Dr D Ogaram
   P.O. Box 4637
   KAMPALA
   Tel: (256 41) 232 680
   or 234 290

Ukraine
Ukrayina

Responsibility	
Main authority responsible for approval/registration of pesticides	1
Main authority responsible for pesticide use directives, permits and training	2, 3, 4
Main authority responsible for establishing national maximum pesticide residue limits	5
Main authority responsible for control of pesticide residues in food	6
Main laboratory for analysis of pesticide residues in food	6
Main laboratory for quality control of pesticide products	7
Main organisation responsible for applied research regarding pesticide use. Information on related regional research	2
Designated national authority for notifications and information regarding banned and severely restricted pesticides including the PIC procedure	-
Codex Alimentarius contact point	6
National IRPTC correspondent	8
Infoterra national focal point	9
National IFCS focal point	8
National association of pesticide manufacturers	10

Responsible organisations

1. State Interdepartmental Commission on Testing and Registration of Plant Protection Products, Growth Regulators and Fertilizers
   9, Suvorova Street
   KIEV 252010
   Tel. (044) 290 0061
   Fax: (044) 290 3284

2. Institute of Plant Protection
   Ukrainian Academy of Agrarian Sciences
   Vasilkovskaja Stret 33
   KIEV 252022
   Tel: (044) 263 1124
   Fax: (044) 263 2185

3. Institute for Occupational Health
   75, Saksagansky Street
   KIEV 252033
   Tel: (044) 220 8030
   Fax: (044) 220 6677

4. National Institute of Health
   Ministry of Health
   6, Heroiv Oborony Street
   KIEV
   Tel: (280 44) 261 4773
   Fax: (280 44) 261 4773

5. Hygienic Regulation Committee
   Ministry of Public Health
   75, Saksagansky Street
   KIEV 252033
   Tel:(044) 220 4366, 220 4777
   Fax: (044) 206 677

6. Ministry of Public Health
   7, Grushevsky Street
   KIEV 252021
   Tel: 293 9417
   Fax: 293 9484

7. Organization on Certification of Mine-Chemical Products (Ekokhimcert)
   98, Stryiskaya Street
   LVYV 290026
   Tel: (0322) 399 669
   Fax: (0322) 344 061, 344 089

8. Prof. Y.I. Kundiev
   Institute for Occupational Health
   75, Saksagansky Street
   KIEV 252033

9. Infoterra/Ukraine
   Attn: Prof. N. Ermoshenko
   Director,
   Ukrainian Institute of Scientific-Technical and Economic Information
   (UKRINTEI)
   Gorky Street, 180
   KIEV 252171
   Tel: (38 44) 268 2522
   Fax: (38 44) 268 2541

10. Ministry of Industry
    34, Kreshchatik Street
    KIEV 252001
    Tel: (044) 226 2623
    Fax: (044) 228 7054

United Arab Emirates

al-Imarat al'-Arabiyah al-Muttahidah

Responsibility	
Main authority responsible for approval/registration of pesticides	1
Main authority responsible for pesticide use directives, permits and training	
Main authority responsible for establishing national maximum pesticide residue limits	-
Main authority responsible for control of pesticide residues in food	3, 4, 5, 6, 7
Main laboratory for analysis of pesticide residues in food	4, 5
Main laboratory for quality control of pesticide products	
Main organisation responsible for applied research regarding pesticide use. Information on related regional research	8
Designated national authority for notifications and information regarding banned and severely restricted pesticides including the PIC procedure	9
Codex Alimentarius contact point	10
National IRPTC correspondent	2
Infoterra national focal point	11
National IFCS focal point	-
National association of pesticide manufacturers	

Responsible organisations

1. Ministry of Agriculture and Fisheries
P.O. Box 1509
DUBAI

2. The Deputy Chairman
Higher Environmental Committee
Ministry of Health
P.O. 1853
DUBAI
Tel: (971 4) 233 021

3. Secretariat General of UAE Municipalities
Chamber of Commerce Building
P.O. Box 3774
ABU DHABI

4. Municipality of Abu Dhabi
Central Laboratory for Food and Public Health
P.O. Box 263
ABU DHABI

5. Municipality of Dubai
Food Control Laboratory
P.O. Box 67
DUBAI

6. Municipality of Sharjah
Central Food Laboratory
P.O. Box 22
SHARJAH

7. Municipapity of Al Ain
Health Section
P.O. Box 1003
AL AIN

8. Agricultural and Animal Products Department
P.O. Box 1004
AL AIN

9. The Director
Plant Protection and Extension Service
Ministry of Agriculture and Fisheries
P.O. Box 1509
DUBAI
Tel: (971 4) 224 136
Fax: (971 4) 232 781

10. Federal Director
Department of Preventive Medicine
Ministry of Health
P.O. Box 848
ABU DHABI

11. The Director General
Infoterra NFP
Federal Environmental Agency
P.O. Box 5951
ABU DHABI
Tel: (971 4) 777 363
Fax: (971 2) 770 501

United Kingdom

United Kingdom of Great Britain and Northern Ireland

Responsibility

Main authority responsible for approval/registration of pesticides	1, 2
Main authority responsible for pesticide use directives, permits and training	1, 2
Main authority responsible for establishing national maximum pesticide residue limits	1, 3
Main authority responsible for control of pesticide residues in food	1, 3, 4, 5
Main laboratory for analysis of pesticide residues in food	6
Main laboratory for quality control of pesticide products	
Main organisation responsible for applied research regarding pesticide use. Information on related regional research	7
Designated national authority for notifications and information regarding banned and severely restricted pesticides including the PIC procedure	11
Codex Alimentarius contact point	12
National IRPTC correspondent	13
Infoterra national focal point	13
National IFCS focal point	15
National association of pesticide manufacturers	14

Note

Poisons information/control centres in Belfast, Birmingham, Edinburgh, Leeds, London, Newcastle and Penarth

Responsible organisations

1. Pesticides Safety Directorate
Mallard House
Kingspool
3, Peasholme Green
YORK YO1 2PX
Tel: (44 1904) 640 500
Fax: (44 1904) 455 733

2. Health and Safety Executive (HSE)
Pesticide Registration Section
Magdalen House
Stanley Precinct
Bootle
MERSEYSIDE L20 3QZ
Tel: (44 151) 951 4000
Fax: (44 151) 951 3317

3. Food Science Division
MAFF
Ergon House
c/o Nobel House
17 Smith Square
LONDON SW1P 3JR
Tel: (44 171) 238 6245

4. Food Standards Division
MAFF
Ergon House
c/o Nobel House
17 Smith Square
LONDON SW1P 3JR
Tel: (44 171) 238 6278

5. Food Safety Directorate
MAFF
Colney Lane
NORWICH NR4 7UQ
Tel: (44 1603) 259 350
Fax: (44 1603) 501 123

6. Laboratory of the Government Chemist
Queens Road
TEDDINGTON
Middlesex TW11 0LY
Tel: (44 181) 943 7572
Fax: (44 181) 943 2767

7. Central Science laboratory
MAFF
Hatching Green
HARPENDEN
Herts AL5 2BD
Tel: (44 1582) 715 241
Fax:(44 1582) 762 178

8. Central Veterinary Laboratory
MAFF
New Haw
WEYBRIDGE
Surrey KT15 3NB
Tel: (44 1932) 341 111
Fax: (44 1932) 347 046

9. Natural Resources Institute
Central Avenue
Chatham Maritime
CHATHAM
Kent ME4 4TB
Tel: (44 1634) 880 088
Fax: (44 1634) 880 066

10. Scottish Agricultural Sciences Agency
East Craigs
EDINBURGH EH12 8NJ
Tel: (44 131) 244 8890
Fax: (44 131) 244 8940

Responsible organisations

11. Department of the
    Environment
    Room A328 Romney House
    43 Marsham Street
    LONDON SW1P 3PY
    Tel: (44 71) 276 8550
    Fax: (44 71) 276 8333

12. Consumer Protection Division
    MAFF
    Ergon House
    c/o Nobel House
    17 Smith Square
    LONDON SW1P 3JR
    Tel: (44 171) 238 6702
    Fax: (44 171) 238 6763

13. Department of the
    Environment
    Attn: Ms Judith Deschamps
    Room P3/185
    2 Marsham Street
    LONDON SW1P 3PY
    Tel: (44 171) 276 5672
    Fax: (44 171) 276 5713

14. British Agrochemicals
    Association (BAA)
    4 Lincoln Court
    PETERBOROUGH PE1 2RP
    Tel: (44 1733) 349 225
    Fax: (44 1733) 62523

15. Department of Health
    Attn: Dr. R Fielder
    Room 634B
    Skipton House
    80 London Road
    LONDON SE1 6LW
    Tel: (44 171) 972 5322
    Fax: (44 171) 972 5156

United States

United States of America

Responsibility	
Main authority responsible for approval/registration of pesticides	1
Main authority responsible for pesticide use directives, permits and training	
Main authority responsible for establishing national maximum pesticide residue limits	1
Main authority responsible for control of pesticide residues in food	2, 3, 6
Main laboratory for analysis of pesticide residues in food	2, 3
Main laboratory for quality control of pesticide products	
Main organisation responsible for applied research regarding pesticide use. Information on related regional research	5
Designated national authority for notifications and information regarding banned and severely restricted pesticides including the PIC procedure	6
Codex Alimentarius contact point	7
National IRPTC correspondent	8
Infoterra national focal point	9
National IFCS focal point	11
National association of pesticide manufacturers	10

Note

Poisons information/control centres in each state

Responsible organisations

1. Registration Division (7505C)
Office of Pesticide Programs
US Environmental Protection Agency
401 M Street S.W.
WASHINGTON D.C. 20460
Tel: (703) 3055447, 308 8000
Fax: (703) 308 8005

2. Food Safety and Inspection Service
US Department of Agriculture (USDA)
WASHINGTON D.C. 20250
Tel: (202) 720 7025

3. Food and Drug Administration
Department of Health and Human Services
5600 Fisher's Lane
ROCKVILLE, MD 20857
Tel: (301) 443 1544

4. Health Effects Division (7509C)
Office of Pesticide Programs
US Environmental Protection Agency
401 M Street S.W.
WASHINGTON D.C. 20460
Tel: (703) 305 7351

5. Agricultural Research Service
US Department of Agriculture
WASHINGTON D.C. 20250
Tel: (202) 720 3656

6. The Assistant Administrator Prevention, Pesticides and Toxic Substances
Environmental Protection Agency
401 M Street S.W.
WASHINGTON D.C. 20460
Tel: (1 202) 260 2902
Fax: (1 202) 260 1847

7. Executive Officer for Codex
Food Safety and Inspection Service (FSIS)
Room 3175, South Building
US Department of Agriculture
WASHINGTON D.C. 20250
Tel: (202) 720 9150
Fax: (202) 720 5124

8. Ms. Linda Traver (7407)
Office of Prevention, Pesticides, and Substances
US Environmental Protection Agency
401 M Street S.W.
WASHINGTON, D.C. 20460
Tel: (202) 260 3938

9. Ms. Emma McNamara
Manager,
US/Infoterra NFP (3404)
Office of Administration and Resources Management
US Environmental Protection Agency
401 M Street S.W.
WASHINGTON D.C. 20460
Tel: (1 202) 260 1522
Fax: (1 202) 260 3923

Responsible organisations

10. American Crop Protection Association (ACPA)
The Madison Building
1156 Fifteenth Street N.W. Suite 400
WASHINGTON D.C. 20005
Tel: (202) 296 1585
Fax: (202) 463 0474

11. Dr Lynn Goldman
Assistant Administrator
Office of Prevention, Pesticides and Toxic Substances (OPPTS)
U.S.Environmental Protection Agency
WASHINGTON; D.C. 20460
Tel: (1 202) 260 2902
Fax: (1 202) 260 1847

Uruguary

República Oriental del Uruguay

Responsibility	
Main authority responsible for approval/registration of pesticides	1
Main authority responsible for pesticide use directives, permits and training	
Main authority responsible for establishing national maximum pesticide residue limits	1, 5, 6
Main authority responsible for control of pesticide residues in food	4, 5, 6, 7
Main laboratory for analysis of pesticide residues in food	8, 9
Main laboratory for quality control of pesticide products	
Main organisation responsible for applied research regarding pesticide use. Information on related regional research	10, 11
Designated national authority for notifications and information regarding banned and severely restricted pesticides including the PIC procedure	13
Codex Alimentarius contact point	2
National IRPTC correspondent	3
Infoterra national focal point	12
National IFCS focal point	-
National association of pesticide manufacturers	

Note

Poisons information/control centre: Centro de Informacion y Asesoramento Toxicologico, Hospital de Clinicas, Avenida Italia s/n Piso 7, 11600 Montevideo, Tel: (598) 2470300, Fax: (598) 2600125 [Head: Pr. Ms. E. Fogel de Korc]

Responsible organisations

1. Ministerio de Ganaderia, Agricultura y Pesca
Dirección General de Servicios Agronómicos
Avda. Millán 4703
MONTEVIDEO
Tel: 392 219
Fax: 392 074

2. Ing. Ruperto E. Long
Presidente,
Laboratorio Tecnológico del Uruguay (LATU)
Avda. Italia 6201
c.c.11500
MONTEVIDEO
Tel: 613 821
Fax: 604 753

3. El Directora,
Division Sustancias Peligrosas
Direccion Nacional de Medio Ambiente (DINAMA)
Attn: Sra. Silvia Aguinaga
Zabala 1427
MONTEVIDEO

4. Dirección de Sanidad Animal
Colonia 892, 6° Piso
MONTEVIDEO
Tel: 397 924, 900 234
Fax: 920 733

5. Servicio de Bromatología
Isla de Flores 1323
MONTEVIDEO
Tel: 902 056, 903 032

6. Intendencia Municipal de Montevideo
Servicio de Salubridad Pública
Eduardo Victor Haedo
MONTEVIDEO

7. Comisión Nacional de Residuos Biológicos en Alimentos de Origen Agropecuario
Colonia 892, 2° Piso
MONTEVIDEO
Tel: 914 320, 920 192

8. Dirección de Laboratorios Veterinarios (DILAVE)
Ruta Brig. Gral. Juan A. Lavalleja Km 18
MONTEVIDEO
Tel: 221 078
Fax: 201 063

9. Laboratorio Tecnológico de Uruguay (LATU)
Galicia 1133
MONTEVIDEO
Tel: 906 386, 984 432, 985 802

10. Instituto Nacional de Investigación Agropecuaria (INIA)
Andes 1365 Piso 12
MONTEVIDEO
Tel: 920 550
Fax: 923 633

11. Centro de Investigaciones Agricolas "Alberto Boerger"
Treinta y Tres 1374
MONTEVIDEO
Tel: 900 448, 906 492, 917 154

12. Lic. Annie Hareau
CNC de Infoterra
Director,
Direccion Nacional de Medio Ambiente
Ministerio de Vivienda, Ordenamiento Territorial y Medio Ambiente
Zabala 1427
Casilla 11.000
MONTEVIDEO
Tel: (598 2) 963 954, 965 133
Fax: (598 2) 965 132

13. Direccion General de Servicios Agricolas
Direccion General de Servicios Ganaderos
Ministerio de Ganaderia, Agricultura y Pesca
Av. Millán 4703
MONTEVIDEO
Tel: (598 5) 392 219
Fax: (598 2) 392 074

Uzbekistan
Ozbekistan Jumhuriyäti

Responsibility	
Main authority responsible for approval/registration of pesticides	1
Main authority responsible for pesticide use directives, permits and training	2, 3, 4
Main authority responsible for establishing national maximum pesticide residue limits	2, 3
Main authority responsible for control of pesticide residues in food	3
Main laboratory for analysis of pesticide residues in food	3
Main laboratory for quality control of pesticide products	2, 3
Main organisation responsible for applied research regarding pesticide use. Information on related regional research	4
Designated national authority for notifications and information regarding banned and severely restricted pesticides including the PIC procedure	2
Codex Alimentarius contact point	-
National IRPTC correspondent	2
Infoterra national focal point	5
National IFCS focal point	-
National association of pesticide manufacturers	6

Responsible organisations

1. Chemical Commission of the Republic of Uzbekistan
   A. Timura, 62
   700000 TASHKENT
   Tel: (371 2) 351 180
   Fax: (371 2) 340 449

2. State Committee for Nature Protection
   A. Kadyri Av. 5-A
   700128 TASHKENT
   Tel: (371 2) 414 923, 414 843
   Fax: (371 2) 413 990

3. Ministry of Public Health
   A. Novoi, 12
   700011 TASHKENT
   Tel: (371 2) 411 624
   Fax; (371 2) 411 641

4. Ministry of Agriculture
   A. Navoi, 4
   700004 TASHKENT
   Tel: (371 2) 412 503
   Fax: (371 2) 410 053

5. Infoterra NFP
   State Fund of Scientific and Technical Information
   State Committee on Science and Technology
   700017 TASHKENT
   Tel: (371 2) 391 688, 394 964
   Fax: (371 2) 394 491

6. The Association of the Enterprises of Chemical Industry
   Lakhuti, 42
   700000 TASHKENT
   Tel: (371 2) 369 261, 363 700
   Fax: (371 2) 335 898

Vanautu

Ripablik blong Vanuatu, République de Vanuatu, Republic of Vanuatu

Responsibility	
Main authority responsible for approval/registration of pesticides	-
Main authority responsible for pesticide use directives, permits and training	
Main authority responsible for establishing national maximum pesticide residue limits	-
Main authority responsible for control of pesticide residues in food	1
Main laboratory for analysis of pesticide residues in food	-
Main laboratory for quality control of pesticide products	
Main organisation responsible for applied research regarding pesticide use. Information on related regional research	-
Designated national authority for notifications and information regarding banned and severely restricted pesticides including the PIC procedure	2
Codex Alimentarius contact point	-
National IRPTC correspondent	3
Infoterra national focal point	3
National IFCS focal point	-
National association of pesticide manufacturers	-

Responsible organisations

1. Director
   Dipatmén blong Helt
   PMB 009
   PORT VILA
   Tel: (678) 22512
   Fax: (678) 24420

2. The Director,
   Department of Agriculture and Horticulture
   Private Mail Bag 040
   PORT VILA
   Tel: (678) 22525
   Fax: (678) 25265

3. Mr. Ernest Bani
   Head,
   Environment Unit
   Private Mail Bag 063
   PORT VILA
   Tel: (678) 25302
   Fax: (678) 23142

Venezuela
República de Venezuela

Responsibility	
Main authority responsible for approval/registration of pesticides	1
Main authority responsible for pesticide use directives, permits and training	
Main authority responsible for establishing national maximum pesticide residue limits	2
Main authority responsible for control of pesticide residues in food	2, 3
Main laboratory for analysis of pesticide residues in food	3, 4, 5, 6, 16
Main laboratory for quality control of pesticide products	
Main organisation responsible for applied research regarding pesticide use. Information on related regional research	9
Designated national authority for notifications and information regarding banned and severely restricted pesticides including the PIC procedure	10
Codex Alimentarius contact point	11
National IRPTC correspondent	12, 13
Infoterra national focal point	14
National IFCS focal point	-
National association of pesticide manufacturers	15

Note

Poisons information/control centres in Barinas, Barquisimeto, Caracas, La Victoria and Valencia

Responsible organisations

1. Ministerio de Sanidad y Asistencia Social
Dirección de Malariología y Saneamiento Ambiental
C.S.B. Edificio Sur piso 1
CARACAS
Tel: 483 2673, 483 3478

2. Ministerio de Sanidad y Asistencia Social
Dirección de Salud Pública
C.S.B. Edificio Sur piso 8
CARACAS
Tel: 483 3533, 483 3881

3. Instituto Nacional de Higiene "Rafael Rangel"
Ciudad Universitaria
CARACAS 1051
Tel: 619 811, 622 6416

4. Fundación Centro de Investigaciones del Estado para la Producción Experimental Agroindustrial (CIEPE)
Zona Industrial San Felipe-Edo
Yaracuy
Apartado 100
CARACAS
Tel: (58 54) 314 551, 413 457, 313 392

5. Fundación Centro de Investigaciones del Estado para la Producción Experimental Agroindustrial (CIEPE)
Avda. Andrés Bello
Edificio Centro Andrés Bello, Torre Este piso 5, Oficina 54
CARACAS
Tel: 781 3423, 781 8219

6. Ministerio de Fomento
Dirección de Industria
Centro Simón Bolivar
C.S.B. Torre Sur piso 6
CARACAS
Tel: 483 2806

7. Ministerio de Sanidad y Asistencia Social
División Higiene de los Alimentos
Centro Simón Bolívar
C.S.B. Edificio Sur piso 3, Oficina 103
CARACAS 1010
Tel: 420 657, 483 4766, 482 4035

8. Universidad Central de Venezuela (UCV)
Facultad de Agronomía
Departamento de Química
Maracay-Estado
Aragua
CARACAS
Tel: (58 43) 460 834, 453 242

9. Universidad Central de Venezuela (UCV)
Fondo Nacional de Investigaciones Científicas (FONAIAP)
Maracay-Estado
Aragua
CARACAS
Tel: 836 965

10. Ministerio del Ambiente y de los Recursos Naturales Renovables
Direccion General Sectorial de Calidad Ambiental
Torre Sur,
Centro Simon Bolivar
Piso 18
CARACAS
Tel: (58 2) 408 1522, 408 1526
Fax: (58 2) 481 0236

11. Dr. D Yanez
Sección de Registro de Alimentos
Ministerio de Sanidad y Asistencia Social
Centro Simón Bolívar,
Edificio Sur, piso 2
CARACAS
Tel: 482 4035, 482 0657

12. Dra. Maritza Rojas
Directora de Investigaciones Toxicas de la Universidad de Carabobo
VALENCIA
Tel: (58 42) 319 279

13. The Director General
Oficina de Educación Ambiental, Desarrollo Profesional y Relaciónes Internacionales
Ministerio del Ambiente y de los Recursos Naturales Renovables (MARNR)
Apartado 6623
1010 CARACAS
Tel: 408 1230, 481 1473
Fax: 483 2445

14. Lic. Ellida Rivas de Carrasquel
Directora de Informatica
Infoterra
Ministerio del Ambiente y de los Recursos Naturales Renovables
Centro Simón Bolívar
Torre Sur, piso 11
CARACAS 1010
Tel: (58 2) 484 4210
Fax: (58 2) 483 9112

15. Asociación de Fabricantes de Productos Químicos Agropecuarios (AFAQUIMA)
Avenida Principal de Macaracuay
Edificio Multicentro Macaracuay
Piso 4 - Oficina No. 1a
Urbanización Macaracuay
CARACAS 1071-A
Tel: (58 2) 256 3652, 256 1284
Fax: (58 2) 256 0418, 256 5504

16. Universidad Central de Venezuela (UCV)
Facultad de Veterinaria
Maracay-Estado
Aragua
CARACAS
Tel: (58 43) 467 472, 464 625

Vietnam
Cong Hoa Xa Hoi Chu Nghia Viet Nam

Responsibility	
Main authority responsible for approval/registration of pesticides	1, 2, 8
Main authority responsible for pesticide use directives, permits and training	
Main authority responsible for establishing national maximum pesticide residue limits	1, 3, 5, 8
Main authority responsible for control of pesticide residues in food	3, 4
Main laboratory for analysis of pesticide residues in food	3
Main laboratory for quality control of pesticide products	2, 3, 4
Main organisation responsible for applied research regarding pesticide use. Information on related regional research	6, 7
Designated national authority for notifications and information regarding banned and severely restricted pesticides including the PIC procedure	10
Codex Alimentarius contact point	5
National IRPTC correspondent	8
Infoterra national focal point	9
National IFCS focal point	-
National association of pesticide manufacturers	-

Note

Poisons information/control centre: 48 Hai Ba Trung Street, Hanoi [Head: Dr D Huy Khac]

Responsible organisations

1. Plant Production and Protection Department (PPPD)
   Ministry of Agriculture and Food Industry
   189B Tay Son Street
   HANOI
   Tel: (84 4) 581 197
   Fax: (84 4) 512 719

2. Department of Products Quality
   Ministry of Agriculture and Rural Development
   1 Bach Thao,
   Ba Dinh
   HANOI
   Tel: (84 4) 232 751
   Fax: (84 4) 234 425

3. Pesticide Quality and Residue Control Center
   189B Tay Son Street
   HANOI
   Tel: (84 4) 513 590
   Fax: (84 4) 521 719

4. Pesticide Quality and Residue Control Center
   Ministry of Agriculture and Rural Development
   28 Mac Dinh Chi Street,
   Distr.1
   HOCHIMINH CITY
   Tel: (84 18) 241 114

5. Vietnam Codex Alimentarius Committee
   Nghia Do, Tu Liem
   HANOI
   Tel: (84 4) 343 407
   Fax: (84 4) 361 771

6. Institute of Epidemiology
   Ministry of Health
   1B Yersin Street
   HANOI
   Tel: (84 4) 254 361
   Fax: (84 4) 212 894

7. Institute of Occupational and Environmental Health
   Ministry of Health
   1 B Yersin Street
   HANOI
   Tel: (84 4) 254 361
   Fax: (84 4) 212 894

8. National Environmental Authority
   Ministry of Science, Technology and Environment (MOSTE)
   39 Tran Hung Dao Street
   HANOI
   Tel: (84 4) 228 752
   Fax: (84 4) 251 518

9. Vietnam Infoterra NFP
   National Centre for Scientific, Technological and Environmental Information and Documentation
   24, Ly Thuong Kiet Street
   HANOI
   Tel: (84 4) 26308
   Fax: (84 4) 263 127

10. Département des sciences et des technologies
    Ministère de l'agriculture et de l'Industrie alimentaire
    Bach Thao
    Quartier Ba Dinh
    HANOI
    Tel: (84 4) 232 751, 237 534
    Fax: (84 4) 234 425

Western Samoa

Independent State of Western Samoa, Malo Sáoloto Tutóatasi o Samoa i Sisifo

Responsibility	
Main authority responsible for approval/registration of pesticides	5
Main authority responsible for pesticide use directives, permits and training	6
Main authority responsible for establishing national maximum pesticide residue limits	1, 5
Main authority responsible for control of pesticide residues in food	6
Main laboratory for analysis of pesticide residues in food	-
Main laboratory for quality control of pesticide products	
Main organisation responsible for applied research regarding pesticide use. Information on related regional research	7
Designated national authority for notifications and information regarding banned and severely restricted pesticides including the PIC procedure	2
Codex Alimentarius contact point	1
National IRPTC correspondent	3
Infoterra national focal point	4
National IFCS focal point	-
National association of pesticide manufacturers	-

Responsible organisations

1. Chief
Public Health Division
Department of Health
P.O. Box 192
APIA
Tel: 21212

2. The Director
Department of Agriculture,
Forests, Fisheries and
Meteorology
P.O. Box 1874
APIA
Tel: 22561, 22564
Fax: 22565

3. The Director-General for
Health
Department of Health
P.O. Box 192
APIA

4. Mr Tepa R. Suaesi
Infoterra NFP
Division of Environment and
Conservation
Department of Lands, Surveys
and Environment
Private Bag
APIA
Tel: (685) 26681
Fax: (685) 23176

5. Pesticides Technical
Committee
c/o The Director
Department of Agriculture,
Forests, Fisheries and
Meteorology
P.O. Box 1874
APIA
Tel: 22561, 22564

6. The Registrar of Pesticides
Department of Agriculture,
Forests, Fisheries and
Meteorology
P.O. Box 1874
APIA
Tel: 22561, 22564

7. Research Division
Department of Agriculture,
Forests, Fisheries and
Meteorology
P.O. Box 206
APIA

Yemen

al-Jumhúríyah al-Yamaníyah

Responsibility	
Main authority responsible for approval/registration of pesticides	1
Main authority responsible for pesticide use directives, permits and training	
Main authority responsible for establishing national maximum pesticide residue limits	-
Main authority responsible for control of pesticide residues in food	2, 3, 4, 5, 6
Main laboratory for analysis of pesticide residues in food	-
Main laboratory for quality control of pesticide products	
Main organisation responsible for applied research regarding pesticide use. Information on related regional research	7, 8
Designated national authority for notifications and information regarding banned and severely restricted pesticides including the PIC procedure	-
Codex Alimentarius contact point	9
National IRPTC correspondent	-
Infoterra national focal point	10
National IFCS focal point	-
National association of pesticide manufacturers	-

Responsible organisations

1. Gessim El-Hasharat
Markaz Abhath El-Kod
Muhafethet Abyan
P.O. Box 1611
ADEN

2. Idarret Sihet Al-Biaá
Khormakssar
Crater
ADEN

3. Shubit El-Muwasefat
Wizarret Al-Sinaá
P.O. Box 300
ADEN
Tel: 51181
Telex: 237

4. Gessim El-Bassatien WEL-Sinaat Al-Githaiah
Markaz Abhath El-Kod
Muhafethet Abyan
P.O. Box 1611
ADEN

5. Mukhtaber Moragabet El-Asmak
Wizarret El-Therwa Al-Samakiah
P.O. Box 1242
Hedjuff
ADEN
Tel: 22687, 22788

6. Mukhtaber Muragabet El-Agthiah
Shará Madram
Al-Má la
P.O. Box 4200
Khormakssar
ADEN

7. Markaz Abhath Siyun
Muhafehet Hadramout
P.O. Box 1611
ADEN

8. Markaz Abhath El-Kod
Muhafethet Abyan
P.O. Box 1611
ADEN

9. General Director for Measurements
Ministry of Economy, Supply and Trade
SANAA

10. Mr. Hussein Al-Gunied
General Secretary, Environment Protection Council
P.O. Box 19719
SANAA
Tel: (967 1) 262350
Fax: (967 1) 215604

Zaire
République du Zaïre

Responsibility	
Main authority responsible for approval/registration of pesticides	1
Main authority responsible for pesticide use directives, permits and training	
Main authority responsible for establishing national maximum pesticide residue limits	1
Main authority responsible for control of pesticide residues in food	6, 7
Main laboratory for analysis of pesticide residues in food	-
Main laboratory for quality control of pesticide products	
Main organisation responsible for applied research regarding pesticide use. Information on related regional research	-
Designated national authority for notifications and information regarding banned and severely restricted pesticides including the PIC procedure	2, 7
Codex Alimentarius contact point	3
National IRPTC correspondent	4
Infoterra national focal point	5
National IFCS focal point	-
National association of pesticide manufacturers	-

Responsible organisations

1. Département de l'Agriculture
Division de la Défense des Végétaux
B.P. 8722
KINSHASA

2. Monsieur le Directeur
Chef de Service des établissements humains et protection de l'environnement
Ministère de l'environnement et tourisme
B.P. 12.348
KINSHASA 1
Tel: 33250, 33251

3. lère Direction des Études et de la Politique Agricoles
Ministère de l'Agriculture et du Développement Rural
B.P. 8722
KINSHASA 1
Tel: 31126

4. Monsieur le Directeur
Département des Affaires Foncières, Environnement et Conservation de la Nature
B.P. 12348
KINSHASA 1

5. Mr. Ntondo Lumuka Nantole
Direction Programmation Formation et Relation Internationales
Ministere de l'Environnement
15, Avenue des Cliniques
B.P. 12348
KINSHASA 1
Tel: (243 12) 33250

6. Département de l'Agriculture
B.P. 8722
KINSHASA

7. Direction de la Production et Protection des Végétaux
Ministère de l'Agriculture
B.P. 8722
KINSHASA 1

Zambia

Republic of Zambia

Responsibility	
Main authority responsible for approval/registration of pesticides	1
Main authority responsible for pesticide use directives, permits and training	1, 8
Main authority responsible for establishing national maximum pesticide residue limits	1, 2, 3, 4
Main authority responsible for control of pesticide residues in food	1, 3, 5
Main laboratory for analysis of pesticide residues in food	4
Main laboratory for quality control of pesticide products	12
Main organisation responsible for applied research regarding pesticide use. Information on related regional research	1, 4, 6, 7, 8
Designated national authority for notifications and information regarding banned and severely restricted pesticides including the PIC procedure	1
Codex Alimentarius contact point	10
National IRPTC correspondent	1
Infoterra national focal point	11
National IFCS focal point	-
National association of pesticide manufacturers	9

Responsible organisations

1. The Director
   Environmental Council of Zambia
   P.O. Box 35131
   LUSAKA
   Tel: (260 1) 224 009, 286 435
   Fax: (260 1) 223 123

2. Ministry of Health
   P.O. Box 30205
   LUSAKA
   Tel: 211 528, 213 435

3. Zambia Bureau of Standards
   P.O. Box 50259
   LUSAKA
   Tel: 251 062

4. Secretary General
   National Council for Scientific Research
   P.O. Box CH 158
   LUSAKA
   Tel: 281 081

5. The Director
   Food and Drug Control Laboratory
   P.O. Box 30138
   LUSAKA
   Tel: 218 477, 218 864

6. Department of Forestry
   P.O. Box 70228
   NDOLA
   Tel: 612 251, 614 524

7. Department of Agriculture
   Mt Makulu Agriculture Research Station
   P.O. Box 7
   CHILANGA
   Tel: 278 008

8. University of Zambia
   P.O. Box 32379
   LUSAKA
   Tel: 213 221, 253 952

9. Zambia Agrochemicals Association
   P.O. Box 31994
   LUSAKA 10101
   Tel: (260 1) 214 325
   Fax: (260 1) 243 910

10. The Secretary
    Food and Drugs Control
    Ministry of Health
    P.O. Box 30205
    LUSAKA

11. Mr. Wycliff Muship
    Information Services Unit
    National Council for Scientific Research
    P.O. Box 310158
    LUSAKA
    Tel: (260 1) 281 081
    Telex: 40005 za

12. The Principal Scientific Officer
    Pest Control Centre
    National Council for Scientific Research
    P.O. Box 49
    CHILANGA
    Tel: 278 655, 278 155

Zimbabwe
Republic of Zimbabwe

Responsibility	
Main authority responsible for approval/registration of pesticides	1
Main authority responsible for pesticide use directives, permits and training	2, 4
Main authority responsible for establishing national maximum pesticide residue limits	2
Main authority responsible for control of pesticide residues in food	1
Main laboratory for analysis of pesticide residues in food	3, 10
Main laboratory for quality control of pesticide products	-
Main organisation responsible for applied research regarding pesticide use. Information on related regional research	4, 10
Designated national authority for notifications and information regarding banned and severely restricted pesticides including the PIC procedure	5, 7
Codex Alimentarius contact point	3
National IRPTC correspondent	7
Infoterra national focal point	8
National IFCS focal point	6
National association of pesticide manufacturers	9

Note

Poisons information/control centre: Drug and Toxicology Information Service, Univesity of Zimbabwe, Medical School, P.O. Box A 178, Avondale, Harare, Tel: (263) 479 1631 (ext. 172), Faz: (263) 473 2828, 479 5019, 479 0233 [Head: Dr (Mrs) O J Kasilo]

Responsible organisations

1. Ministry of Agriculture
Causeway
HARARE

2. Ministry of Health
Causeway
P.O. Box CY 1122
HARARE

3. The Government Analyst
The Government Analyst Laboratory
P.O. Box 8042
Causeway
HARARE
Tel: 792026

4. Ministry of Agriculture
Research and Specialist Service
Chemistry and Soil Research Institute
P.O. Box 8100
Causeway
HARARE

5. The Director
Department of Research and Specialist Services
P.O. Box 8108
5th Stret Extension
Causeway
HARARE
Tel: (263 4) 704 531

6. The Director,
Directorate of the Environmental Health Department
Ministry of Health and Child Welfare
P.O. Box CY 1122
Causeway
HARARE
Tel: (263 4) 730 011
Fax: (263 4) 793 634

7. Hazardous Substances and Articles Control Office
Ministry of Health
P.O. Box 8204
Causeway
HARARE
Tel: (263 4) 730 011

8. Mrs. Sithabiso Gandure
Department of Natural Resources
Makombe Complex, Block 1
Causeway
HARARE
Tel: (263 4) 705 661
Fax: (263 4) 793 123

9. Agricultural Chemicals Industry Association (ACIA)
c/o Bayer Zimbabwe PVT
P.O. Box AY78
Amby
HARARE
Tel: (263 4) 48464
Fax: (263 4) 48455

Responsible organisations

10. Ministry of Agriculture
    Research and Specialist
    Services
    Plant Protection Research
    Institute
    P.O. Box 8100
    Causeway
    HARARE
    Tel: 704 531
    Fax: 728 317